APPLIED Fishery Science

Volume 1

S.M. Shafi

ATLANTIC
PUBLISHERS & DISTRIBUTORS

Published by
ATLANTIC PUBLISHERS AND DISTRIBUTORS
B-2, Vishal Enclave, Opp. Rajouri Garden, New Delhi-27
Phones : 25413460, 25429987, 25466842

Sales Office
7/22, Ansari Road, Darya Ganj, New Delhi-02
Phones : 23273880, 23285873, 23280451
Fax : 91-11-23285873
web : www.atlanticbooks.com
e-mail : info@atlanticbooks.com

Vol. I ISBN 81-269-0277-9
Vol. II ISBN 81-269-0289-2
Set ISBN 81-269-0290-6

Typeset at
APD Computer Graphics, Delhi
Printed in India at
Mehra Offset Press, Delhi

PREFACE

The present book *Applied Fishery Science* is the outcome of efforts made by me for as long as five years to bring the universally spread documented informations or data, both traditional as well as research oriented recent findings, before the scientific world in a consolidated form, specially before those who are concerned with fish farming and for capture of fishes from varied water sources. It is hoped that the academicians, researchers/fishery scientists and post-graduate students of fishery science will cherish the author's endeavour in finding appreciable utility of the book.

Several valuable informations have been provided on the fish farming requisites, such as possible ways/methods of the quality fish seed procurement, their safe transportation, design and layout for the construction of new ponds, maintenance of required water quality in the ponds, availability of choiced food items at proper times, use of right kinds of fertilizers and provision of supplementary feeds at emerging times. Several result-oriented informations have been provided under separate headings in the book.

Infrastructural details and operation methods of fishing gears and some informations on related accessaries, particularly trawlers, fishing boats and vessels and several kinds of crafts, traps and angling implements to catch sporting fishes have been given with lucid illustrations. Presentation of recent data on fish catches (year-wise in most cases) has obviously enhanced the quality of book.

Matters, on several specific topics like hilsa culture, trout culture, live-fish culture, cage culture, paddy-cum-fish culture, integrated fish culture, brackish water fish culture, prawn culture, shrimp culture, molluscan fisheries, recent aquaculture practices and useful seaweeds and algal products, have been beautifully described in the book. Trade and commerce prospects with the inclusion of accounts of fish processing, fish preservation, various kinds of fish by-products and scope of marketing of fish have been made highly explicit.

A fine attempt has been made to provide as many as 51 aspects of applied Fishery science, merging together in one book in order to meet the requirements of sylabi of Universities, technical/ professional institutions, advanced centres of fisheries education and research.

I am thankful to Prof. S.N. Ahsan (Vice-Chancellor, Patna University), Dr. R.K. Sharan (Patna University), Dr. K.A. Jafri (Aligarh Muslim University), Dr. N.C. Datta (Calcutta University) and several other eminent zoologists for their benevolence of giving expert advice from time to time and sending useful literature without delay, whenever the need arose.

Suggestions, if any, for the improvement of this book in the next edition shall be thankfully acknowledged.

S.M. SHAFI

CONTENTS

Preface *iii-iv*

VOLUME 1

1. General Introduction 1-9
2. Culturable Fishes 10-21
3. Natural Food and Feeds 22-43
4. Natural Breeding 44-52
5. Induced Breeding 53-69
6. Fish Seeds 70-82
7. Transportation of Fish Seeds 83-91
8. Fish Farm Construction 92-100
9. Quality of Water Media and Bottom Soil 101-116
10. Nursery Pond Management : Fry Production 117-128
11. Rearing Pond Management : Fingerling Production 129-134
12. Stock Pond Management : Fish Production 135-148
13. Pond Fertilization 149-158
14. Synthetic Feeds 159-168
15. Carp Culture 169-181
16. Hilsa Culture 182-188
17. Trout Culture 189-196
18. Live Fish Culture 197-207
19. Cage Culture 208-213
20. Paddy-cum-Fish Culture 214-218
21. Integrated Fish Farming 219-225
22. Fish Culture in Brackish Water Ponds 226-236
23. Prawn Culture in Fresh Water 237-242

24. Prawn Culture in Brackish Water 243-257
25. Shrimp Culture 258-268
26. Cold Water Fisheries 269-278
27. Riverine Hydrobiological Features 279-292
28. Riverine Fisheries 293-306

VOLUME 2

29. Effects of Dam Construction on Fisheries 307-316
30. Main Features and Management of Reservoirs 317-327
31. Reservoir Fisheries 328-342
32. Sewage-fed Pond Fisheries 343-353
33. Estuarine Water Fisheries 354-366
34. Hydrobiological Features of Seas 367-375
35. Off-Shore/Deep Sea Water Fisheries 376-381
36. Coastal Fisheries 382-393
37. Molluscan Fisheries 394-406
38. Fishing Gears of Inland Water Sources 407-431
39. Gears and Vessels for Sea Fishing 432-447
40. Fish Migration 448-462
41. Pollution Hazards to Fishes 463-472
42. Fish Diseases and Control 473-492
43. Control of Weeds 493-509
44. Control of Unwanted Fishes and Insects 510-521
45. Nutritive Value of Fish 522-532
46. Spoilage 533-540
47. Fish Preservation and Processing 541-552
48. Fish By-Products 553-571
49. Fish Marketing 572-578
50. Recent Aquaculture Practices 579-590
51. Useful Seaweeds and Algal Products 591-601

1

General Introduction

Inland fish culture has vast scope in India because a large number of village ponds, tanks, reservoirs, brackish water areas, rivers, irrigation channels are available for raising fish population. According to latest survey work of 1995, the country has 28.55 lakh ha ponds and tanks, 20.50 lakh ha reservoirs, 171334 km. of rivers and canals and 14.22 lakh ha of brackish water resources (IIM Report, 1995). The present annual fish production of 8.4 million tonnes includes the major contribution (4.5 million tonnes) from inland fisheries. These figures of fish production are, however, not attractive when the figures of other countries, particularly south-eastern Asian countries, are taken into consideration.

Improvement in fish production figure can be achieved by intensive culture programmes and adoption of latest technologies developed by various institutions in the country. As the scope of increase of fish production from marine sources is limited, being mainly based on operation by latest gear technology, major efforts are concentrated on aquaculture. Through scientific management and latest fish production technologies, it is envisaged to obtain fish production to the extent of even 11 tonnes/ha/yr from village ponds and tanks against the average production of less than 1 tonne/ha/yr. Although this target appears superficially exaggerated, positive instance is that the fish farmers in West Bengal, through culture in properly managed village ponds, have been harvesting a production of 2,000-3,500 kg/ha/yr, which fetches a profit margin of Rs. 30,000 per hectare in current valuation. Likewise, through fish based mixed farming systems (with prawn), farmers earn a lot, fetching valuation of upto Rs. 60,000/ha/yr.

Fish culture in ponds and profitable scope of fisheries from reservoirs, brackish waters need more attention under proper management technology. However, the requirement for getting more catches from coastal deep sea waters is of different nature, the real needs being the development of infrastructure of fishing gears, motorised fishing boats, vessels, etc., so that more catches can be brought to shore. In addition to these, there is need of latest technology for preservation of fish and other aquaculture products, efficient marketing system and better strategies for development of trade and commerce including export business. In consideration to the present status of fish production and envisaged potentialities of various sources of water for fish production, there is acute need of new target-based planning and sufficient means for execution of the plan-works, whatsoever.

Fish Culture in Ponds. Carp culture is gaining popularity in India as there is now more attention by the state and central governments concerned towards execution of development programmes. Carps, all of which are scale-fishes, belong to the family cyprinidae. Incidentally most of the popular cultivable species of fish belong to this family. In composite culture practices, three Indian species of carps (*e.g.* catla, rohu, mrigal) and three exotic carps, *viz.* silver carp, grass carp and common carp are preferred over other carps in consideration to utilization of available food in various strata of water.

One of the major problems in cultivating the aforesaid carps is the non-availability of proper quality of seeds of these species for ready stocking in fish ponds, as they (except the case of common carp) do not breed in confined waters even though they mature in these water bodies. The yield of seed by common carp in ponds is of poor quality due to wild breeding. To overcome this problem a technique of controlled spawning has been developed. The major carps generally breed in suitable riverine habitats during the monsoon months and the fertilized eggs, hatchlings and larvae (together known as spawn) drift down the rivers along with the water-current. In the plain areas of Northern India, particularly in the river Ganges, one can observe, during monsoon months, the fishermen on boats collecting spawn and fry of carps in large numbers with the help of specially designed nets. The fish seeds thus collected are transported to markets usually in the earthen containers and aluminium hundies. There is well established and flourishing trade of fish seeds at Kolkata in West Bengal.

With the application of modern management techniques, evolved indigenously, and also with the availability of vast water sources, substantial increase in the fish production can be procured. Know how of pond culture and reservoir fisheries has boosted for enhancing the fresh water fish production quantum. About 9.47 million hectares of water is under fish culture and inland fish capture at present out of 2.25 million hectares available in India as ponds and tanks and 2.09 million hectares as lakes and reservoirs. Presently India produces hardly 1.3 million tonnes of fish from fish farming *i.e.* culture fisheries, predominantly from freshwater water bodies. However, some brackish water fisheries are still, by and large, 'Capture Oriented.' These figures of low rate of fish production clearly indicate the degree of non-utilization and under-utilization of the readily available cultivable waters. The relevance of reasons may be several, such as (i) shortage of critical inputs, (ii) ineffectiveness of the functioning of fishermen co-operatives and Gram Panchayats, (iii) prevalent short term lease system for fishing rights, etc.

Fish culture practices may be affected by several external factors, such as the occurrence of pollution, disease, flood condition, the inadequate water supply, presence of weeds, algal bloom condition, and the presence of predacious organisms in ponds, particularly in nursery ponds. If we sort out the existing constraints, *i.e.*, the lapses involved in the culture practices, it will be realised that most of these are directly because of poor management or management failures in the execution of norms and adoption of cautions. Some other reasons are of purely socio-economical nature which need thorough attention of sorting out the existing constrains and evolving some better techniques to do away with the hurdles.

Reservoir Fisheries. India has a vast inland fishery water sources in the form of ponds, tanks, lakes, reservoirs and rivers, as already mentioned before. Reservoirs are mainly constructed for irrigation, hydel-power projects and for control of floods. However, some natural lakes are also available for irrigation purposes as well as fisheries. In most cases the reservoirs, thus created for various purposes, are utilized effectively for fish production as well. Reservoir fishery essentially represents stocking-cum-fish production system. There are about 975 reservoirs in the country, of which 550 are major reservoirs. In a recent study, conducted by the Indian Institute of Management, Ahmedabad on Inland Fish Marketing in India, it is identified that 419 major reservoirs, spread in 11 states with a total area of 12,05,946 ha are available in the country, which can be used for fish production, but unfortunately the present figure of quantum of fish production from reservoir, in most cases, is noticeably low. The average productivity of a large reservoir is only 5-10 kg/ha. However, in the case of some small reservoirs, productivity as high as 40-80 kg/ha has also been reported, and higher yields upto 80 kg/ha/yr. have been recorded under experiment authority. In the case of large number of reservoirs, the management authority is generally not attentive towards fulfilling the basic needs of the reservoir for fishery. In the absence of any significant stocking of fish seed in them, the present reservoir productivity can be termed, as per experts opinion, wild productivity.

The fishery activity in the reservoir starts usually after 3-5 years of its impoundment. It should be stocked with fish seeds of economic fast growing carps to explore thc vast resource. For an ideal management of reservoir, the hydrobiological conditions of reservoirs, the basic characteristics of the catchment area of the reservoir and finally information on the fish fauna present in the reservoir are to be ascertained. Stocking of fish seeds of proper quality is the next important step in the maintenance of reservoir fishery. A reservoir harbours a number of fish species comprising major carps, exotic carps, minor carps and minnows, *siluroidae* (*e.g.* catfishes), clupeids (mostly *Gadusia onapara*) and many other varieties of fish species. Different fish species are known to breed at different preferred places and time/season. Indian and exotic major carps are the most valuable fish species for culture in ponds and reservoirs, but normally they do not spawn there. The lacustrine condition of the reservoirs also prevents them from breeding in the reservoirs. For maintaining an optimum level of productivity in the reservoir and to make the best use of it in future also, it is essential to stock ponds and reservoirs with calculated amounts of fish seed of the desired species at a suitable period of time. October and November are usually the best months for stocking because advanced fingerlings in good condition are generally available at this time.

Generally 500-800 fingerlings of Indian major carps measuring 50-100 mm are good enough for one hectare water surface of a reservoir. Considering the feeding habits of the major carps, it is recommended to stock catla, rohu and mrigal in the ratio of 3:6:1. However stocking rate as well as stocking ratio may change depending upon the provision of artificial feeding in the reservoir. Thus for major reservoirs of more than 100 acres, about 200 fingerlings/hectare water surface is the optimum requirement; for medium sized reservoir of 500-1000 ha in area about 500 fingerlings/ha and for minor reservoirs of less than 500 ha in area, 100 fingerlings/ha area are considered as ideal stocking ratio.

Natural Spawning. For natural spawning of the Indian and Chinese carps, the essential requirement is optimum level of hydrological and climatic conditions. Fish receiving hormonal injection spawn only when the temperature is within optimum ranges, *i.e.*, between 24° – 30°C. The other requirement is related to sufficient dilution of pondwater by rain, cool rainy days, fairly wide range of pH and dissolved oxygen level in water. Flowing river water is more conducive for spawning. Fishery experts of Taiwan and USSR also lay emphasis on the above mentioned climatic features and hydrobiological conditions of waters.

Induced Breeding. The success achieved in the use of fish pituitary hormones on spawning cultivated species of fishes, particularly major carps, has played a vital role in popularising the fish aquaculture concept all over the country. In recent decades much attention has been given towards expansion of the fish culture practices by procurement of pure fish seeds after induced breeding. Even in China, Taiwan, Indonesia and other South-eastern countries, the problem of shortage of fish seeds has been solved by the introduction of hypophysation technique. Besides major carps, grey mullets, milk fish and a few other important salt water fishes are cultured in ponds in several countries. Whatsoever, the fish seeds are in great demand for culture also in brackish water and some reservoirs in India. In Russia, the use of hypophysation technique is widely emphasized for the conservation of sturgeon in rivers, as the construction of dam across rivers had brought constrains in the procurement of seed from the riverine sources *i.e.,* by natural spawning method. The efforts of fishery scientists are even now more concentrated in improving technique for the procurement of quality seeds in plenty. The design and operation methods of hatcheria are related aspects in which the scientists have great scope to suggest for improvement in the technology.

Brackish Water Fishery. The brackish water sources of the country can be profitably organised for prawn fish culture. It has been estimated that, through scientific management, a net profit of Rs. 20,000-30,000/ha/yr. could be obtained from brackish water fish culture practices. This technology, if adopted fully for brackish water culture, can provide self-employment. A large number of youths can find gainful employment in seed collection, production and transport to meet the ever increasing demand for prawn seed. According to one estimation, the brackish water aquaculture alone has potentiality for generating employment for about five million people in the country.

Constraints in fish culture practices and remedies. In fish culture practices, control of aquatic weeds, predaceous organisms and fish diseases assures great significance in view of extent of hindrances in activities of fishes and incumbent threat to their life. For a real success in fish culture practices, the nursery pond should be cleared of weeds, miscellaneous (*e.g.* predaceous) fishes, aquatic insects and other enemies of young fry. Eradication of the aquatic insects, particularly the backswimmers (notonectids), is absolutely essential because they take a heavy toll of the carp spawn. This operation should be done a day prior to the stocking of prawn. For control of backswimmers, soil-oil emulsion is spread on the surface of pondwater, 12 to 14 hours before stocking. There are many other methods including the use of insecticides, but the fry population may not be spared fully from the adverse effect. So, the detergent-oil is more proper device to control aquatic insects. It seems that there is scope of further investigations, *i.e.*, experimental studies to explore novish ideas to control above mentioned constrains.

Containment of aquatic weeds in the cultivable ponds becomes one of the primary problems in the overall management of culture practices. Though it is a global problem, it is more acute in the tropical and subtropical countries. Their excessive growth reduces the productivity of the water body, as they utilize nutrients, obstruct the penetration of sunlight into the water body by shading, reduce the eutrophic zone, restrict the living space for fish, impede fishing operations and, in the long run, lead to gradual silting up of the water body. The weeds also provide cover for predatory and weed fishes which breed profusely and overpopulate in the water body. This results in the reduction of available food to the cultural species. Above all these, they may threaten the survivality of the fish and other organisms either by oxygen depletion or liberation of toxic substances after their decomposition. Though several methods of weed control (manual, mechanical, chemical and biological) are known, there is open avenue even now to find out some new strategy to solve this gigantic problem.

Discharge of factory wastes or industrial waste effluents into rivers may pollute rivers and adversely affect the riverine fisheries. Besides this possibility, there arises chance of large scale mortality of fish and fish fry in rivers due to access of highly toxic insecticides/pesticides, widely used in agriculture practices. In consideration to this adverse situation, there is scope of exploration of proper remedial measures or finding out some suitable way towards abatement of pollution.

Construction of dams across many rivers in India has restricted the movement of migratory fishes considerably and has brought about changes in the environment and natural breeding ground of many fishes, thus depleting the fishery prospects in these waters. With the envisaged possibility of construction of more dams in future, the condition will be still worse.

Natural calamities, brought about by floods almost every year in most of rivers in Northern India are affecting the fish seed collection prospects from rivers to a great extent. The embankments constructed along many rivers to prevent floods, has also destroyed the natural spawning ground of carps.

Considering implications of above mentioned nature and magnitude, it is apprehended that in future the scope of river fisheries will be gradually depleted, if suitable remedial measures were not taken. Consequently, the chances of producing sufficient quantity of fish seed from the riverine source, as prevalent now, will be greatly reduced. The need for depending on seed production by induced breeding method, thus, appears to be all the more important. Further improvement in the fish breeding technique and rearing of fish can compensate or bring relief to envisaged adversities by natural or man-made calamities.

Fish Diseases. Like other animals, fishes also suffer from diseases, with the introduction of high yielding technologies of intensive culture, which suggests the use of fertilizers, supplementary feeds of many kinds, the incidence of disease has gone very high, and enormous sufferage is experienced by the culturists or fish farmers. The diseased fish generally shows unusual movements and has an unhealthy appearance. The body skin may have a film-like covering, blisters, lesions, swellings or spottings. The gill cover may become pale and the gills may also have white or red spots on their filaments.

It is not unusual to come across the reports of massive mortality of fish in some reservoirs or in a certain portion of a water body like a river, canal, sea coast, etc. Sometimes such a phenomenon is simply the result of improper application of the technologies of intensive fish culture and the poor level of farm management. The outbreaking disease is facilitated by several factors, such as the pollution of water, the presence of excessive quantities of organic matters in water due to heavy manuring in ponds and the excess of feeding of fish, the stagnation of water which change its gaseous composition and the reduction of oxygen content. Nevertheless, high density of fish population inhibiting normal growth and development of fish facilitates fast spread of diseases among fishes. It has been found that scaleless fishes *e.g.* catfish are more susceptible to diseases than the scaled fishes, *e.g.* carps. Whatsoever the causes, symptoms developed, preventive measures suggested and chemical treatments applied, further scope of research work is open on this aspect even now. The application of modern technology strategies to culturable pond should be with caution. Expert advice should always be sought for any odd in the fish culture practices.

Fishing Gear Technology. General success in fishing, particularly in the exploitation of reservoirs, is not at a satisfactory level even now. At present the water bodies are, in most cases, characterised by poor fish output : 1 tonne fish obtained from 1 km stretch of river or about 10 kg fish from 1 hectare reservoir. The foremost reason for this is the underestimated stocking of fingerlings, unsatisfactory quality of these seeds, and the poor management set-up. Secondly, the quality of fishing gears and operative methods for harvesting fish are not appropriate. For an efficient management of a water body, it is necessary to have the desired fish of the required size only at a proper time, which is possible only by using the right type of fishing gear. Hence correct fishing gear is of great importance in increasing fish production. Further scope of investigation lies in the improvement of fishing gear technology.

All above mentioned aspects, discussed in brief above, invite further investigational approach to find out ways and means for overall improvement in the inland fisheries. One must admit that a huge fish catch from reservoir or cultivable pond is the ultimate aim of all developmental activities undertaken in the inland fisheries.

2

Culturable Fishes

The main criteria for selection of fish for culture in culturable pond water are, in order of priority, as follows :

1. Fish should grow to big size with enough flesh in a short period of time.
2. Fish should have efficiency to utilize and convert the organic production in water as fish flesh.
3. Fish should be capable of directly making use of the primary organisms in the water in which they live. Such fish are considered more suitable than those which depend upon micro-organisms or other small organisms (animals/fishes).

Most commonly acknowledged cultural species in India are six major carps which are preferred over others in consideration to their highly desirable cultural qualities. Three of these carps have Indian Origin (*i.e.* are endogenous species), but the rest three are exotic species, now adapted well in Indian conditions of waters. These are :

1. Endogenous species, *e.g.*, *Catla catla* (commonly called catla), *Labeo rohita* (called rohu), *Cirrhinus mrigala* (called mrigala).
2. Exogenous Species, *e.g.*, *Hypophthalmichthys molitrix* (called silver carp), *Ctenopharyngodon idella* (called grass carp) and *Cyprinus carpio communis* (called common carp).

These fishes get outright selection on merit basis. The main or foremost criteria of their selection is the choice of food by these fishes in respect to their kinds available and their place of availability in the pond water. A brief mention of the pond water zone preference can satisfy the urge by any concerned person.

Catla and silver carp are both surface feeders, but the silver carp, not the catla specimens, alone is able to effectively utilize the phytoplankton which is so abundant in pond water. It has been on record that both these species attain good growth, reaching table size within less than a year in normal conditions of waters and with optimum rate of stocking (*e.g.* density).

Mrigal and common carp are bottom feeders, whereas rohu is a column feeder. Grass carp is a valuable addition in the choice of carps as it feeds on aquatic vegetation. It does not compete for food with other carps because of specialized feeding habits.

Though competition is least among carps for the utilization of pond and they have almost classified feeding habits, there is always chance of overlapping in choice of food, as noticed in some cases during gut analysis. Whatsoever, the combination of these carps for culture enables utilization of most of food organisms present in pond.

In China and Indonesia, fish culture practice includes zooplankton-feeding big-head carp, *Aristichthys nobilis,* along with the phytoplankton-feeding silver carps, *Hypophthalmichthys molitrix*, and the weed-eating grass carp, *Ctenopharyngodon idella.* In India, in fish culture practice in some places, the detritus-feeder major carp, calbasu (*Labeo calbasu)* and gonius (*Labeo gonius*) are included in appreciation of their feeding habits and attainment of their growth.

There is variation in the growth rate of different species of fish and, also, the individuals of some species of fish, flourishing in different culture ponds, may exhibit variation in the growth achieved. The climate and ecological conditions of habitats have impact on the growth rate and maturity attainment by fishes. Also quantum of availability of consumable food items in a pond brings direct effect on the growth. The consensus of opinion is that the fish growth and crop production mainly depend on the management practices, sufficiency of oxygen, abundance of natural food and perfection in general hygienic condition of pond.

Some basic information on the feeding habits, growth and maturity of different culturable carps is given below.

Catla. This carp is the fastest growing indigenous major carp (Fig. 2.1). It normally matures within 2 years in the plains of northern India, but more commonly in the third year. However, males of normal catla and of its golden coloured mutant variety have been sometimes observed in oozing condition within one year also. The size (total length) attained by the fully grown adult fish varies between 19 to 28 cm. The head portion is broad, and the greatest width of the head equals its length behind the middle of the eye. The dorsal profile is more convex than that of the abdomen. The colour is greyish above, becoming silvery along the side and beneath, and the fins are dark black. It is commonly available throughout fresh waters of northern India.

The size at maturity varies considerably from habitat to habitat. With proper supplementary feeding, the size of the 3-year old fish (*i.e.*, at maturity) may range between 3 and 6 kg. The males carry weight less than the females in the same age group. Sometimes in the specimens of catla, when fed with excessive diet or in certain ecological conditions, the females do not mature fully and possess excessive fat deposit which interferes with the gonad development.

Rohu. This is another fast growing major carp which is found in abundance in the freshwaters of Northern India, Pakistan, Bangladesh and Myanmar, and is universally considered as one of the most delicious fish of these areas (Fig. 2.2). It is largely a column feeder fish, feeding on a wide variety of algal forms as well as decayed macrovegetation. The growth rate of this fish is slightly lower than that of catla. The fish attains maturity in 2 years, but the males may mature even in 1 year. The mature fish of 2 years weighs 1.5-2.0 kg and has a length of 24-26 cm. The dorsal profile of the body is more convex than the abdomen and the colour is bluish black along the back, becoming reddish black along the sides and silvery beneath. There is red mark on each scale, surrounded by dark margin. However, the fins are black.

Mrigal. This is a bottom dweller fish subsisting mainly on semi-rotten vegetable matters and detritus and available in abundance in the freshwaters of Northern part of India, Pakistan, and

Myanmar (Fig. 2.3). The adult fish is 16-26 cm in length and the colour is silvery dark grey along the back and whitish below, whereas their eyes are conspicuously golden. It attains maturity in 2 years and, when well fed, it weighs 1.5 to 2.5 kg.

Silver Carp. This fish is exotic, being introduced in India in 1959. Of all major carps, this is the fastest growing carp, now available throughout freshwaters of Northern India (Fig. 2.4). Like catla this fish is a surface feeder, but while catla feeds mainly on zooplanktons, the silver carp feeds predominantly on phytoplanktons. The close-meshed nature of the gills of this fish puts it in an advantageous position over catla, the fastest growing Indian major carp. As both phytoplankton and zooplankton food items are available in the surface water, there is competition of food between these two species.

The fish generally matures fully in 2 years, but usually the males attain maturity in a year. The fish at 2 years of age weighs 2.5-5 kg. This wide variation in weight is related to the availability of plenty of food and the proper maintenance of pond..

Grass Carp. This is another fast growing exotic major carp (Fig. 2.5). It is a voracious feeder on certain types of freshwater macrovegetation. As a component of the composite culture fish, it occupies its own food zone and does not compete with other species. It also contributes substantially in the form of semi-digested excreta which are directly consumed by the bottom feeders, such as common carps. The decayed half consumed vegetable matter also serves as a manure for the plankton production.

The fish generally matures in 2 years, but the males may attain maturity earlier, that is even in one year, showing milt-oozing condition. The ripe female grass carp of the male age is usually larger in size.The brood fish may attain an approximate size of 2-4 kg in 2 years (Fig. 2.5).

Common Carp. This represents as a domesticated exotic carp (*Cyprinus carpio var communis*), commonly available in the freshwaters of northern India. This is actually the common representative of the Indian strain of the original common carp, *C. carpio var communis* (Fig. 2.6). This is an omnivorous scavenger fish generally feeding the soft bottom organisms at the bottom of ponds margin.

The common carp shows rapid growth and matures even in 6 months, but generally good class of brood fish can be obtained from 1 to 2 year old stock. The mature brood fish attains 1-2 kg body weight within a year. A unique feature is that the weight of testis in relation to the body weight is usually higher in common carp than in all other carps mentioned before.

Maturation in Carps

Fishes possessing developed gonads (ovary in the case of female and testis in male) with full ripe ova or spermatory sacs are known as mature fishes. Age and size of first maturity in carps vary considerably as per nutrient and ecological conditions of the pond or reservoir in which they live.

Female carps generally mature within two years, except catla and common carp, which take 3 years and 1 year respectively for reaching maturity. On the other hand, males of all other carps attain maturity within 1 year.

In gonads of both sexes marked cyclic morphological and histological changes occur before attaining full maturity. On the basis of shape, size, colour of the gonads (ovary or testis) and some other histomorphological characteristics, seven maturity stages are recognized, as mentioned below. It has been observed that the time of maturation of gonads is not uniform in all carps or even in all species of the same carp. It may be faster or slower, depending on the age, type and quality of food, daily photoperiod, *i.e.*, day length and environmental temperature. The ratio of development of the gonads is slow till the third stage of their maturatior, but the development of the ovary afterwards is very fast, and the stages of development are completed within less time. At stage-V, there is conspicuous gain in the weight of the ovary, attaining 20-30% of the total body weight. However, the gain in weight of the male is generally only 5-10%.

The seven recognized stages in gonadal development of the carps are the following :

(I) Immature (II) Early Maturing (III) Late Maturing (IV) Maturing (V) Mature (VI) Spawning (VII) Spent.

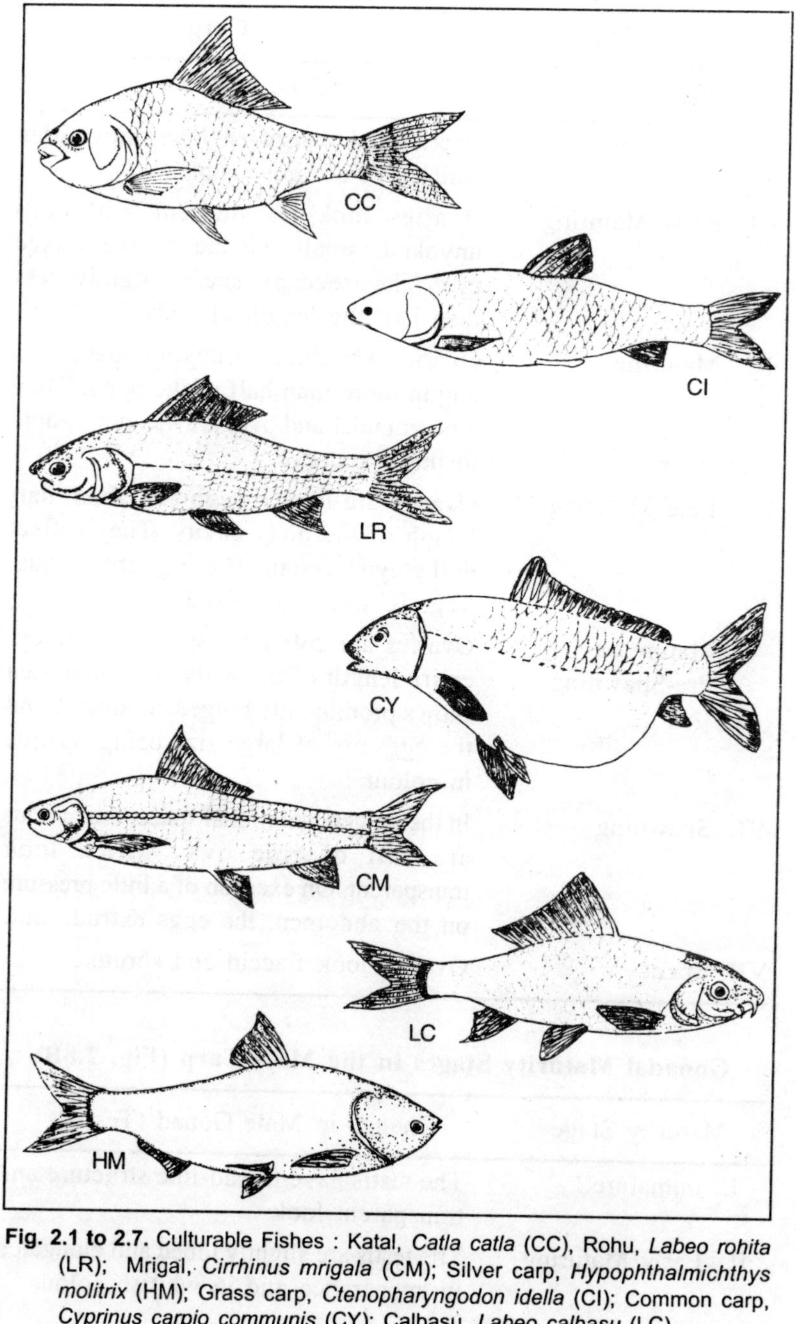

Fig. 2.1 to 2.7. Culturable Fishes : Katal, *Catla catla* (CC), Rohu, *Labeo rohita* (LR); Mrigal, *Cirrhinus mrigala* (CM); Silver carp, *Hypophthalmichthys molitrix* (HM); Grass carp, *Ctenopharyngodon idella* (CI); Common carp, *Cyprinus carpio communis* (CY); Calbasu, *Labeo calbasu* (LC).

Gonadal Maturity Stages in the Female Carp (Fig. 2.8A)

Maturity Stage	Ovarian Features
I. Immature	Ovaries look transparent long strips within which ova are not visible.
II. Early Maturing	Ovaries look transluscent with-eggs unyoked, small, visible to the naked eye. They occupy space slightly less than half the length of body.
III. Maturing	Ovaries are long, occupying space in a length more than half of the body. They look granular and ovas are opaque, visible to the naked eye.
IV. Late Maturing	Ovaries are large, occupying about half length of the body cavity. They reflect dull greyish colour. The eggs are opaque, large in size and yolked.
V. Mature or Pre-Spawning	Ovaries are coloured, spread in almost entire length of the body. The abdomen shows prominently bulged condition, and the eggs are of large size being opaque in colour.
VI. Spawning	In the bulged abdominal space, the ovaries are full of ripe ovas which look transparent. On exertion of a little pressure on the abdomen, the eggs extrude out.
VII. Spent	Ovaries look flaccid and shrunken.

Gonadal Maturity Stages in the Male Carp (Fig. 2.8B)

Maturity Stage	Change in Male Gonad (Testis)
I. Immature	The testis have thread-like structure and transparent look.
II. Early Maturing	The testis are slightly lobed and elongated in appearance and in whitish colour.

III. Maturing	The testes have opaque look, and the main body shows white strip-like appearance.
IV. Late Maturing	The testes have milky colour and convoluted appearance. The fish shows oozing condition, releasing whitish fluid on applying little pressure on the body.
V. Pre-Spawning or Mature	The testes have cream colour and exhibits convoluted condition. The fish shows oozing of milt on applying some pressure on the abdomen.
VI. Spawning	The testes show highly convoluted appearance. The abdomen shows coming out of milt on applying a little pressure on the abdomen.
VII. Spent	The testes look reduced in size, and reflect reabsorption condition.

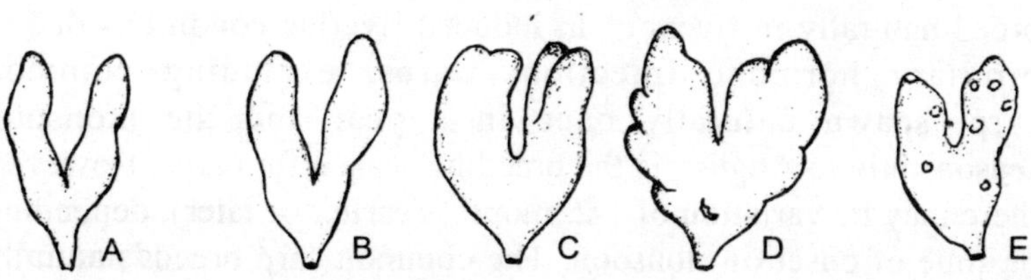

Fig. 2.8A. Five developmental stages of ovary of a major carp: Immature (A) Maturing (B) Mature or prespawning (C) Spawning (D) Spent (E).

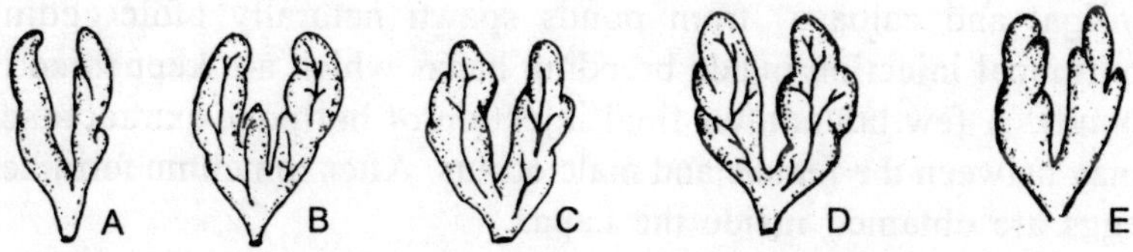

Fig 2.8B. Five developmental stages of testis of a major carp : Immature (A) Maturing (B) Pre-spawning or mature (C) Spawning or Mature II (D), Spent (E).

All carps do not get opportunity to spawn. In those female carps which fail to spawn, regression of gonads or resorption of ovarian eggs occurs at the end of breeding season. The oocytes, failing to release eggs, undergo serious changes and are known as atreotic follicles. The enclosed eggs lose their shape; the vitelline membrane breaks at various positions and, after more disintegration, finally disappears. The yolk droplets loose normal consistency, liquidify so that a continuous mass of yolk is formed. It is finally absorbed in the stroma of the ovary. For carps, it takes time to re-equip themselves and enter a fresh ovarian cycle next season (Chaudhuri *et al.* 1966; Raina, 1976).

Fecundity, which denotes the egg laying capacity of a fish, is fairly high in carps. Carps in general do not exhibit parental care. Fecundity rate is highest in *L. rohita,* next in *C. mrigala* and next in *C. carpio.* Silver carp and grass carp have lowest fecundity.

Spawning

In natural conditions, the above mentioned major carps do not breed in confined waters (ponds or tanks or reservoirs). They breed naturally in rivers or in induced riverine conditions due to pituitary hormone injection. Carps, excluding common carp, spawn naturally once in a year and the monsoon season, July to August, is the breeding season for carps. However, there may be variation of 1-2 months (earlier or later), depending on time of onset of monsoon. The common carp breeds naturally in ponds twice a year in monsoon (July-August) and post-winter (March-April) periods (Alikunhi *et al.* 1964; Singh *et al.* 1972).

The ripe brood fishes of major Indian carps (*e.g.* catla, rohu, mrigal and calbasu) from ponds spawn naturally after getting hormonal injection inside breeding hapas which are kept fixed in ponds. A few hours after final injection of hormone extract a sex play between the female and male occurs. After sometime fertilized eggs are obtained inside the hapa.

Details of informations have been given ahead about spawning by hypophysation, stripping action and artificial fertilization of eggs, induced breeding of common carp.

India has annual fish production of 11 lakh tonnes from the inland resources. This limit of fish production is insufficient in view of increasing number of consumers with the increase of population figure. Increase in fish production can be attained by practising intensive culture programmes. Along with the monoculture, composite culture or polyculture and prawn culture, there is need of expanding scope of culture of other groups of fishes, such as live fish culture, cold water fish culture, fish culture in brackish water, etc.

Conditions for choice of culturable fishes. There are several choices of selection of fish for fish culture practices. These are mainly based on following factors :

(i) Capacity to grow to big size with fleshy body in a short period of time.

(ii) Efficiency to utilize and convert organic production in water into fish flesh. Direct consumer of primary organisms in the water in which they live in come in this list.

(iii) Adaptability to live and breed in different types of water bodies.

(iv) Amiability in behaviour to live together with other species, *i.e.*, not compete each other for food.

(v) Suitability in using food and share with other fishes.

(vi) Acceptable taste of fish.

(vii) Easier availability of fish seeds.

In the polyculture or composite culture practices, the combination of several species is done in consideration to above mentioned factors. However, in the monoculture practice, utilization of the natural food from only one zone is possible. Therefore, a group of fishes in reasonable grounds are selected so as to make use of all food available in the different zones of water. A combination of the surface feedery catla and silver carp and the bottom feeders like mrigal and common carp (omnivorous bottom-feeder), column feeders rohu and vegetation feeder, grass carp shows ideal combination. In some places calbasu (*Labeo calbasu*) or *Labeo gonius* is also considered for inclusion (Fig. 2.9, 2.10). It is essentially, a herbivorous fish, subsisting mostly on algae, diatoms, decayed plant matters and also zooplankton. There is no marked difference in the feeding habits of the fingerlings, juveniles and adults. However in West Bengal, fishermen in some areas introduce some of the less carnivorous fishes such as *Channa striatus* and *Anabas testudineus*. Due attention is given by the fishermen to their introduction only after the carp fry have grown into a fairly large size so that they can escape on any attempt by these mild carnivores to make them their prey. These carnivores generally utilize insects of the pond water as food. Their chase of the carp fry leads to considerable degree of exercise which turns helpful to stipulate growth.

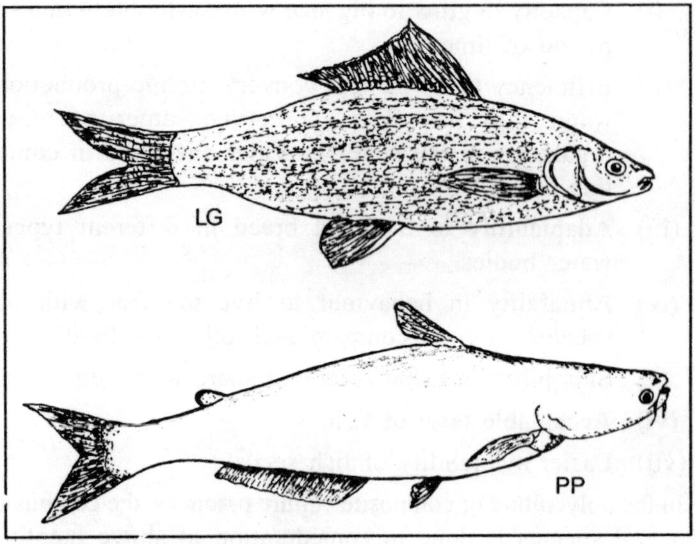

Fig 2.9-2.10. Additional culturable fishes : Kursa bata, *Labeo gonius* (LG); Payas, *Pangasius pangasius* (PP).

There are several carnivorous fishes having essential qualities for inclusion in culture practices in ponds. They are able to show rapid growth, have palatable flesh and can breed in pond unlike Indian major carps. Cat fishes like *Mystus seenghala* and *Mystus oax,* pangash or payas (*Pangasius pangasius*), pabda (*Ompok bimaculatus, O. pabda*) and wallah (*Wallago attu*) are important carnivorous fishes showing suitability for cultural purposes in ponds. The only negative point in their consideration is their predatory food habits. Whatsoever, most valuable point is that they have good market demand because of the palatable flesh.

Pangash or payas feeds mostly on gastropods. It has been found that the fish farmers of Bangladesh prefer to introduce a few pangash fish in the carp culture pond. Their motive behind this selection is to control the population growth of gastropods and, thus, eradicate helminthes which are liable to parasitize fish and cause loss to culture fish. The gastropods act as vectors for helminth parasites.

Another aspect of consideration is that for releasing seeds in culture ponds, the fry and fingerlings are easily available in tanks, paddy fields, irrigation channels and rivers. Their easy availability facilitates cultural practices. Moreover, pure strain of seeds are now available due to induced breeding of catfishes by hypophysation, which is now successfully done.

Very recently culture practices with cat fishes have gained popularity in Eastern India, in view of their flesh quality and their marketing demand in foreign countries like U.S.A., Bahrain, and Australia. In many Southeastern countries, there has been tremendous growth in culture practices with carnivorous fish and this has turned to be a profitable industry, as elaborated ahead in the chapter 'Live-Fish Culture.'

REFERENCES

Ali Kunhi, K.H. and Parmeshwaran, S. 1963a. Induced spawning of the Chinese carps, *Ctenopharyngodon idella* and *hypophthalmichthys molitrix* (C & V) in ponds at Cuttack, India. Curr. Sci. 32 : 103-26.

—— and Sukumaran, K.K. 1962. Preliminary Observations on Chinese carps in India. Proc. Ind. Acad. Sci. 60B (3) : 171-88.

Chen, F.Y., Chow, M. and Sim, B.K. 1969. Induced spawning of the three major Chinese carps in Malacca, Malaysia. Malay. agric. J. 47 : 211-38.

Chaudhuri, H. 1967. Breeding and selection of cultivated warm water fishes in Asia and the Far East. FAO Fish Rep. 44 (4) : 30-36.

Chondar, S.L. 1970. Handbook of Breeding of Indian major carps by Pituitary Hormone. 100p Agra. Satish Book Enterprise.

Raina, H.R. Seasonal Histological changes in the ovary of *Schizothorax esocinus* (Haeckel). Matsya, 21 : 66-71.

Singh, S.B., Sukumaran, K.K., Chakraborty, P.C. and Bagchi, M.M.M. 1972. Observations on composite culture of exotic carps. J. Inland Fish Soc., India 4 : 38-50.

Sinha, V.R.P., Jhingran, V.S. and Ganpati, S.V. 1974. A review of spawning of the Indian major carps. Arch. Hydrobiol. 73 (4) : 518-36.

3

Natural Food and Feeds

In intensive fish culture practices, the main aim is to get maximum fish yield. To achieve this target, sufficient knowledge of the optimum qualitative and quantitative dietary requirement of the organisms being cultured is indispensable. This will be of great help to get maximum yield and to counteract a number of diseases caused by nutritional deficiencies. Furthermore, application of modern techniques relating to monoculture, composite culture, polyculture or mixed culture practices is an essential requirement for enhancing fish production from inland water sources.

Fishes are known to consume various kinds of organisms present in the environment, and convert them to flesh containing mostly protein. However, the natural food present in the environment may not suffice the needs of the growing fish population. Addition of the manures and fertilizers to the water media contributes to flourishing of the planktons which are readily used by plankton feeder fishes such as carps or larval and juvenile stages of most freshwater fishes. The growing young ones, however, need the provision of the supplementary feed as well to retain full growth. It is known that for any species of fish, food habits change with the seasons, life history stages, and the kind of food available.

Considering all these aspects, it is required that the nutritional needs of the fries and fingerling forms are ascertained first and provisions made afterwards for addition of supplementary feeds in proper time.

In vast water areas like those of reservoirs and rivers, the natural food is the only source for fulfilling nutritional requirements of fishes and latter are known to use a wide variety of food items which include the animal matters, plant matters, decaying organic matters (detritus), planktons and other micro-organisms as living matters. Some of them subsist on plant matters only, while others use animal matters as their food. However, a good number of them feed on both animal and plant matters, *i.e.*, have undertaken omnivorous feeding habits. Some of these fishes use planktons (zoo- or phytoplanktons) as their food, but there is less number of such types of fishes. A further case of specialization occurs in few fishes which depend on decaying organic materials formed at the bed of column of water, usually intermixing with mud.

In consideration to this type of food used, fishes belong to herbivorous, carnivorous, omnivorous, plankton-feeder, and detritus feeder groups. However, from the gut analysis, it appears that very few fishes are exclusively herbivorous or carnivorous types as in most such cases some percentage of animal matters (in herbivorous types) or plant matters (in carnivorous types) is usually present in the gut.

BIOLOGICAL REQUIREMENTS OF FISH PONDS

A fish pond accommodates biotic communities of plants and animals which live together and are functionally interrelated in many ways. The green plants make use of the sunlight and, by photosynthesis, transform it into chemical energy. The chemical energy resulting from photosynthesis is utilized by all organisms in the ponds : the producers, consumers and decomposers (Fig. 3.1A, 3.1B). The energy is passed from primary organism to consumer in food chains (Fig. 3.2A, 3.2B). The example for this is as given below :

Chemical energy from algae ⟶ Zooplankton ⟶ Fish.

The functioning of the living communities in the food ecosystem is dependent upon the physical condition of the pond, such as temperature, depth, retentivity of water, pH, dissolved carbon dioxide, dissolved salts, dissolved organic materials and turbidity. Many of these factors are of direct concern to the life of living communities, and are also partly determined by the activities of the biotic communities. Therefore, one can say that, in ponds, there exists a closely interrelated system (Fig. 3.2A, B, C, D).

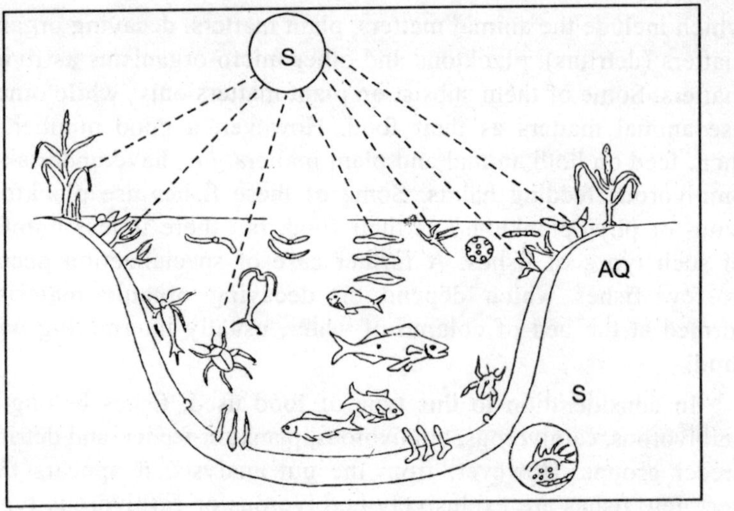

Fig. 3.1A. Ecosystem in a freshwater pond : Sun (S), Aquatic plant (AQ), Decomposers (bacteria, fungi) (D).

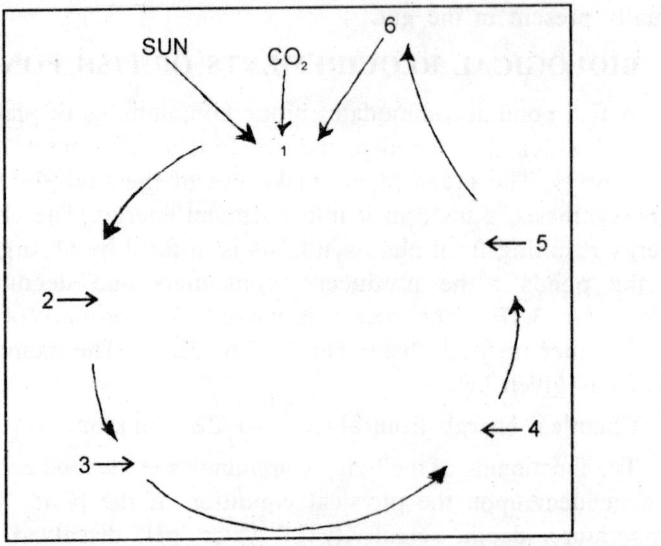

Fig. 3.1B. Food chain in a pond: Legends given with figure. 1: Primary producers, e.g. phytoplankton, autotrophic plants, 2 = Primary consumers e.g. herbivores, crustaceans, molluscs, insects 3 = primary carnivores e.g. small fish, 4: Secondary carnivores e.g. large fish like wallago, 5: Decomposers (e.g. bacteria causing death and decay, 6 = Nutrient salts).

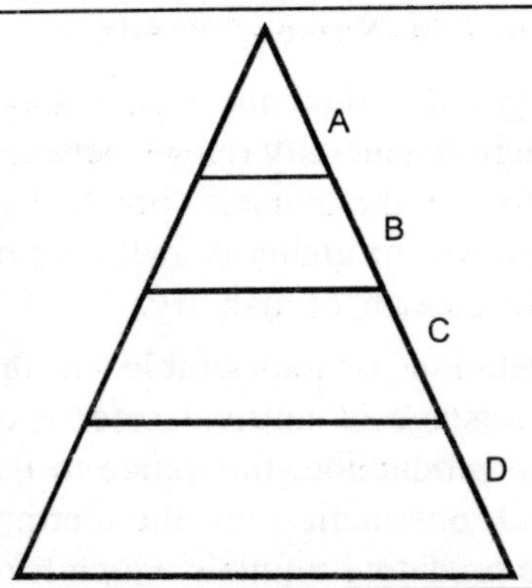

3.2A. Pyramid of numbers in a pond : Carnivore e.g. Large fish (A), smaller fishes (B), copepods (C), Diatoms, Algae (D).

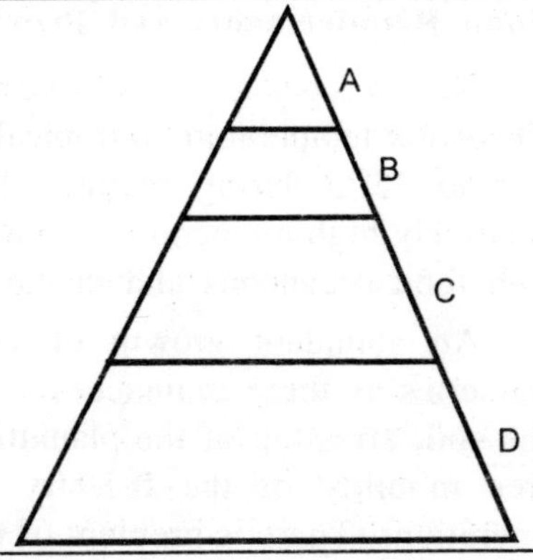

3.2B. Pyramid of energy in a pond : Tertiary consumers (A), Secondary consumers (B), Primary consumers (C), Producers (D).

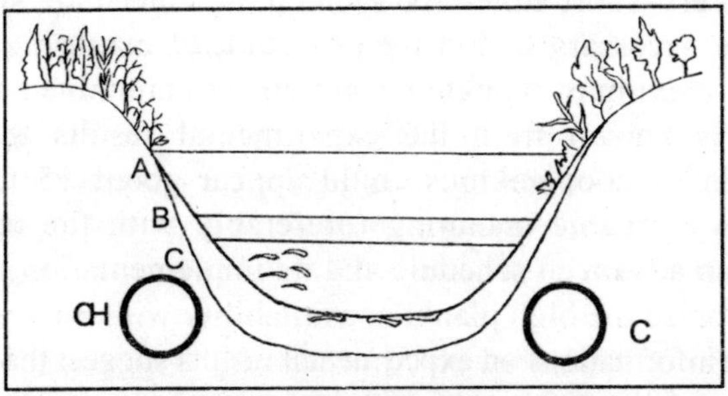

3.2C. Thermal stratification of a pond : Epilimnion I (E_I), Epilimnion II (E_{II}), Hypolimnion (H), Chironomus (C), Chaoborus (CH).

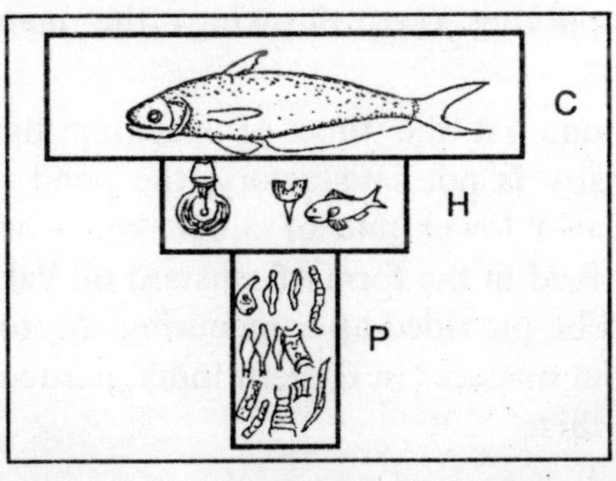

3.2D. Pyramid of a biomass in a pond ecosystem : Carnivore (C), Herbivores (H), Producers (P).

Food Requirements and Provisions in Nursery Ponds

Nurseries being shallow have no problem of thermal stratification. The water temperature in tropical region generally ranges between 25° to 35°C during rearing of fry in the nursery pond. The relatively high temperature is conducive for higher growth rate of fish food organisms and metabolic growth of fish fry.

An abundant growth of vegetation is undesirable in the nurseries as there is unnecessary wastage of nutrient reserve of the soil, arresting of the planktonic production, hindrance to the free mobility of the fish fry and obstruction to the netting operations. There is problem of the predatory animals which take shelter there and eat upon them. As such, there is need of eradication of both weeds and predators.

In prepared nursery ponds, after stocking of spawn or fries, the existing food *i.e.*, available food reserve, generally gets used up in a very short time. As such, it is a need of situation to undertake necessary action for provision of manuring in such a way that maximum planktonic growth is maintained for a long time. It is known from the experimental results that a high population of zooplanktons could appear about 15 to 20 days after heavy organic manuring (preferably with the cow dung). As such an advanced schedule and its implementation is required to synchronise the high plankton availability with the fry stocking. Available informations on experimental results suggest that manuring at a rate of 12,000 to 15,000 kgs. per hectare of the water surface area of pond gives satisfactory result. Manuring should be done by applying (*i.e.,* throwing) the manure along the marginal waters.

If it is found, at the time of stocking the fry, that the planktonic density is not satisfactory, the pond is to be further fertilized, but at a lower rate of 5,000 kg/ha and later on the supplementary feed in the form of mustard oil-cake or groundnut oil-cake should be provided after manuring. A common schedule maintained about manures in Eastern India, particularly Bihar and Orissa, is like this.

Manure Quantity/ha	Manuring item/ha water surface area	Timing for use
10,000-15,000 kg 10,000 kg	Raw cow dung	15-20 days before stocking, if the first manuring did not yield satisfactory result.
200-400 kg	Mustard oil cake	15-20 days before stocking
300-500 kg	Groundnut oil cake	15-20 days before stocking

In Maharashtra the schedule of manuring is generally maintained in four stages (I to IV), one after another, as given below :

Stage	Manure Quantity/ha Fertilizer	Manuring item/ha water surface area	Timing of use
I	700 kg	Raw cow dung	1 day before stocking
	700 kg	Oil cake	
	150 kg	Superphosphate of lime	
	150 kg	Triple phosphate of lime	
II	88 kg	Raw cow dung	1 day before stocking
	350 kg	Oil cake	
III	44 kg 175 kg	Raw cow dung Oil cake	On 2nd day after stocking
IV	22 kg 88 kg	Raw cow dung Oil cake	On any date between 3rd to 10h day, following the date of stocking

If it is found that the pond in question has acidic soil and pH is lower, there should be treatment with lime @ 200-250 kg/ha. This should be done before the treatment with the fertilizers (as listed above). Another suggestion is that the lime treatment may be combined with the fertilizers. However, caution is to be maintained about the treatment with the inorganic fertilizer, specially about its amount, because the inorganic fertilizer induces profuse growth of the phytoplanktonic organisms, which is, unfortunately, an unnecessary blessing due to the general preference of the carp fry for zooplanktons. Moreover, dissolved oxygen in the pond water shows wide fluctuations, sometimes reaching acute depletion with the result that large scale mortality of fry may occur.

Excess of phytoplankton is the main causative factor for this. Zooplanktons in the nursery ponds generally are copeponds, cladocerans, infusoria and rotifera. Fish fries consume these items voraciously (George, 1964).

Supplementary Feeding

The food requirements of fry are judged on the basis of growth co-efficient. The latter is calculated as per undergiven formula :

$$\frac{\textit{The quantity of feed}}{\textit{Weight increase in flesh}}$$

The choice of supplementary feed should be dependent on (i) availability of ingredients in abundance, (ii) transportation facility, (iii) less deterioration quality, and (iv) low cost involvement.

In different areas of the country, several combinations of organic substances have been suggested and tried to fulfil the requirement, but the artificial feed consisting of grain bran (of wheat or rice) and powdered vegetable oil cake (*e.g.* mustard or groundnut oil cake mixed after powdering in equal weights) have been found to give most satisfactory result. Care is to be taken about the extent of powdering. The powder should be so fine that its particles can be easily consumed by the spawn of early fry. The daily ration to be given can be estimated on the basis of fry population and their advancing growth at that time. Considering the average weight of each fry as 0.0014 gram, the total quantity of ration to be given on that date should be like this :

Stocking days	Ration to be given	
	K.H. Alikunhi's (1957) suggestion	S.L. Hora and K.C. Pillay's (1962) suggestion
1st 5 days	Double the weight	Equal to the weight
2nd 5 days	Three times weight	Double weight
3rd 5 days	Four times weight	Three times weight

Estimation of the total weight of the fry is done as per undergiven formula :

Av. no. of stocked fry × av. wt. of a fry/spawn.

The powder should be broadcasted on the surface of water, preferably in the early morning. It has been generally found that all the feed powder introduced in this way is not utilized by the early spawn and fry. So, large quantity drops down in water, reaches the bottom of the pond and gets decomposed. This situation leads to stimulation of the planktonic production. In this way an additional advantage is available by using supplementary feed.

Among Indian carps, mrigal utilizes the artificial feed best, and catla young ones have little choice of utilizing this sort of artificial feed. Among exotic carps, the common carp is the best utilizer.

Another formulation of supplementary feed for feeding growing fry in nursery ponds is a mixture of dry finely powdered notonecta, small prawns and cheap pulses in the ratio of 5:3:2. The rate of broadcasting of the feed is the same as mentioned before for the mixture of the grain bran and vegetable oil cake.

Food Provisions in Rearing Fish Pond

In the rearing pond, which is larger than nurseries but smaller than stocking ponds, the fry or advanced fry is reared to fingerling stage before stocking the latter in the stocking ponds. The main reason for this intermediate rearing is that the food and feeding habits of the fingerling are different from both fry stage and the adult form. In this situation, the transfer of fry to rearing pond is the call of time, *i.e.*, is need based.

The management practices for rearing pond are, in principle, the same as in the case of nursery ponds. The time taken for the rearing of the fry to the fingerling or juvenile stage is usually 2 to 3 months and sometimes even 3-6 months, depending on the growth rate of the young one, which is very much related to the availability of food materials.

The initial requirement is proper manuring of the pond so that production of plankton (both phytoplanktons & zooplanktons) can be enough later on *i.e.*, after stocking of the fingerling or juveniles in the pond) to suffice the requirement of the growing population. The most accepted procedure of manuring, stage by stage, is as follows :

(1) Organic manuring with raw cow dung should be done 8 days prior to stocking of fries.

(2) A second manuring, *i.e.*, this time inorganic manuring, should be done after a month of initial manuring, with the inorganic manures comprising ammonium sulphate, single superphosphate and calcium ammonium nitrate in the ratio of 11:5:1. The rate of application of manure should be 1000-1500 kg per hectare per year. However, the manure should be applied in instalments, *i.e.*, not at a time. If the rearing is to be done for 3 months only, the application of 500-750 kg per hectare inorganic manuring in the small dosage in a number of instalments yields better result.

Due to cheaper and easy availability of organic manure in the form of cow dung in India, both organic and inorganic manures are recommended for rearing ponds. However, in other countries, only inorganic fertilizer is used in the rearing pond.

Kinds of Planktons. Regarding kinds of planktons growing after manuring, there is a long list of species occurring in rearing/stocking ponds of different places in the country (Alikunhi, 1952; George, 1964; Chakrabarty *et al.*, 1979). The list for abundance of most commonly found genera in India is given below.

Phytoplanktons (Fig. 3.3)

Chlamydomonas,	Chlorella	Closterium	Navicula,
Pennularia,	Eudorina	Scenedesmus,	Pediastrum,
Fragilaria			

Zooplanktons (Fig. 3.4)

Diaptomus	Moina	Nauplius	Keratella
Brachionus	Cyclops	Daphnia	

Supplementary Feeding

The fries belonging to fingerling size are fed with the grain bran, *i.e.*, rice/wheat bran and oil cake, *i.e.*, mustard oil cake or groundnut oil cake in the proportion of 1 : 1 by weight daily. The amount to be given is as follows :

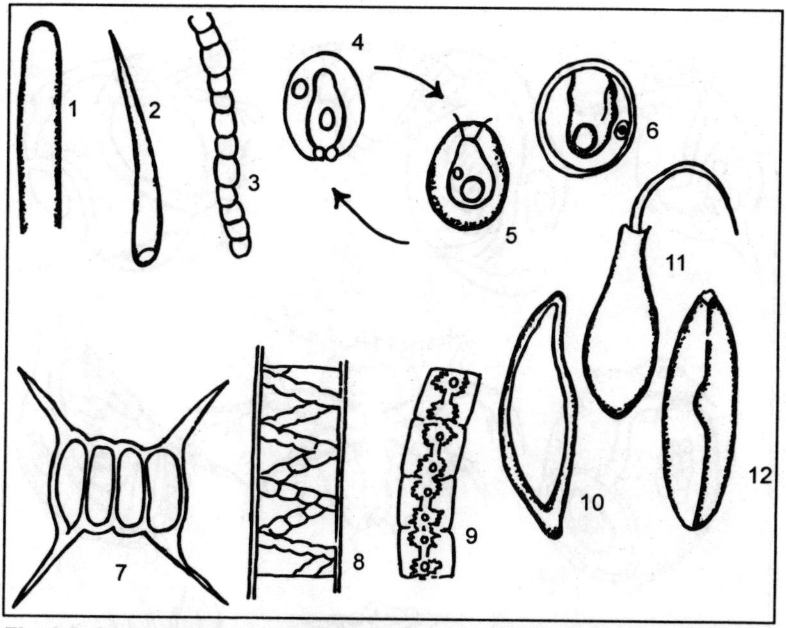

Fig. 3.3. Some common phytoplanktons in a pond: Oscillatoria (1), Anabaena (2), Nostoc (3), Chlamydomonas (4), Eudorina (5), Chlorella (6), Scenedesmus (7), Spirogyra (8), Zygonema (9), Closterium (10), Euglena (11), Pinnularia (12).

(i) During the first month; the feed weight equivalent is the initial weight of the fry.

(ii) During the second or third month : double the equivalent weight of the fry stocked.

The average size and weight attained by different major carps (which are universally recognized as culture species) at the end of 6 months rearing and feeding in pond is given here as per finding by Sinha and others (1973).

Length/Weight attained	Growth figure in major carps					
	Catla	Rohu	Mrigal	Silver Carp	Grass Carp	Common Carp
Length (mm)	277	296 210	366	352	608	295 220
Weight (gram)	334.1	386.8 137.0	571.4	492.2	2530	605.6 248.8

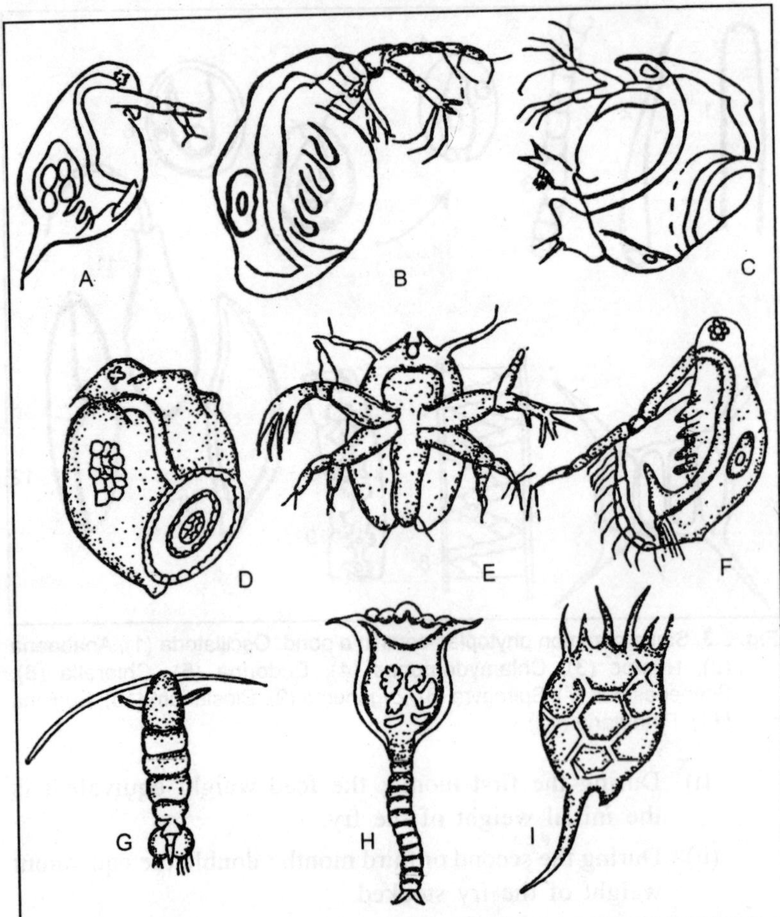

Fig. 3.4. Some common zooplankton in a pond (A), Daphnia (A), Moina (B), Simocephalus (C), Ceratodaphnia (D), Nauplius (E), Diaphanosoma (F), Diaptomus (G), Brachionus (H), Keratella (I).

Rearing and Feeding of Grass Carp and Silver Carp

The pond is stocked with organic manure (*e.g.* cow dung) @ 2850 kg/ha and inorganic fertilizers (*e.g.* ammonium sulphate, single superphosphate, calcium ammonium nitrate) in the ratio of 11:1 @ 700 kg/ha in two equal instalments prior to stocking and, later, two weeks after stocking. The growing young ones are fed with grain bran oil cake in the proportion of 1:1

@ 9 kg/ha/day, and duck weeds (*Lemna)* @ 400 kg/ha occasionally. If the stocking density in the pond was higher than the usual, the ratio schedule should be increased to 12.5 kg/ha/day of grain bran and oil cake mixture (1:1 ratio) and 600 kg/ha of duck weeds (*Lemna*).

The application of these supplementary feeds results in satisfactory growth of the fingerlings. However, the growth rate of the grass carp was not encouraging as this growth was dependent on the availability of weeds in the pond. It is a remarkable fact that when the grass carp cross 250-300 mm size, they turn phytophagous. Thus it is the proper time when the transfer of the young one to the stock pond is necessitated.

Food Provisions in Stock Pond

The main aim of management of stock pond is to obtain maximum fish yield of marketable size in short time. For this objective, there is maximum utilization of all living space and food available in the pond. Several methods of fish culture are practised out of which composite culture with three indigenous and three exotic carps, as mentioned before, has gained highest popularity, as the component fishes make full utilization of available food and attain maximum growth in a shorter period and, nevertheless, more profit margin is available to the fish farmers.

Before stocking operation with fingerlings or juveniles of carps, necessary attention is required towards fertilization of the pond with added nutrients. This endeavour influences for enriching planktonic population in water in the form of phytoplankton or zooplankton. Several species of these flourish in the pond water; however, their specifications show variation in various inland water bodies. Most commonly found genera/species are as follows :

Phytoplankton genera (Fig. 3.3). Green Algae : Microspora, Oedogonium, Pediastrum, Spirogyra, Ulothrix, Pleurococcus, Tribonema.

Blue Green Algae : Anabaena, Merismopedia, Microcystis, Nostoc, Oscillatoria, Rivularia, Tetrapedia.

Zooplankton genera (Fig. 3.4). Cladocerans, Moina, Daphnia, Alona, Ceratodaphnia, Bosmina, and Diaphanosoma.

Rotifers : Brachionus, Keratella, Triarthra, Asplanchna, Rotifer, Polyarthra.

Copepods : Cyclops, Diaptomus, Nauplii, Copepodid stages of several other copepods.

Ostracods : Cypris, Cypridopsis

Flagellates : Euglena

Protozoans : Arcella

The nutritive value of these natural food organisms are undoubtedly of high order, but their availability in natural condition, *i.e.,* without external use of manures/fertilizers throughout the year and covering the growth period of the carps is insufficient or uncertain. The account of nutritive value with other relevant informations as percentage figures, are given below for some natural and artificial foods.

Nutritive value *i.e.* (Chemical Composition) of some natural and artificial foods (in per cent) :

	Components of Food Items						References
	Protein	Carbo-hydrate	Fat	Fibre	Ash	Moisture	
Zooplankton	45-65	10-25	4-8	—	5-20	65-80	MA.V. Lakshmanan *et al.* 1967
Notonectids	56.2	24.1	4.2	—	15.5	—	
Prawn	18.8	32.8	8.4	—	40.0	—	

For a pond with highly clayey bottom soil, having organic deposits at the bottom layer, the use of inorganic fertilizer is highly recommended. For a pond with soil bottom, neither too clayey nor two sandy and, at the same time, with minimum organic content, a combination of raw cow dung plus vegetable oil cake or groundnut oil cake is effective; however, in ponds with acidic soil and low pH, a pretreatment with quick lime (CaO) at the rate of 250-600 kg/ha will be advantageous.

It needs recalling here that the fertilization programme should be put to execution about 2-3 weeks before stocking of fingerlings.

During the post-stocking period, manuring operation is taken up in a number of instalments.

The organic manure that is, cow dung should be applied @ 20-25 tonnes/ha/year in 7 to 11 instalments. The inorganic fertilizers of ammonium salphate, single sugar phosphate and calcium ammonium nitrate in the ratio of 11:5:1 should be mixed together and applied @ 1500-1750kg/ha/year in 4-10 instalments. A relevant and pertinent recommendation is that the organic manure when applied after introduction of the young fish to the stocking pond, should not be ordinarily broadcasted, rather should be put in baskets and placed under water in the corners of the pond. The other alternative is that the manure should be directly pumped under water in the corners of the pond.

Sharing of Natural Food Available in the Stock Pond by Carps

In the carp culture, there is no place for carnivores and in the stock ponds, a mixture of plankton feeders and macrophyte feeders is stocked. Both groups of fish utilize the planktonic communities of the pond to fullest extent, and attain desirable growth. There is least competition among them for getting food, rather they have good sense of adjustment and share foods from either surface, or mid-column or bottom layers, *i.e.*, from all zones of pond water.

In consideration to these qualities of adjustment by the phytoplankton feeder and zooplankton feeder fishes, the fisheries experts have highly recommended for practising multispecies culture in India. For composite culture practices, the major carps selected are the three Indian major carps, *i.e.,* catla, rohu and mrigal, three exotic major carps, that is, two chinese carps (grass carp and silver carp) and the common carp of Bangkok strain (*Cyprinus carpio*). The juveniles of major carps share planktonic food of stocking pond, that is consume plankton efficiently. They exhibit tolerance to one another and utilize the food zones of the ponds most efficiently. All these activities lead to expectancy of a high yield of fishes from culture ponds on harvesting.

In respect to the feeding habits of the grass carp, one valuable additional information is available. The experimental work on composite culture by Sinha and Gupta (1973) have shown that the grass carp attains conspicuously high growth rate with adequate supply of weed and the result is further promising, when the weed was given on wooden platform in the middle of pond, rather than developing them in water (Fig. 3.5). The fish (grass carp) grasps the food easily with least involvement of energy. The fish farmer can from the weed stock left on the platform, judge the feeding requirement of the fish and act upon accordingly.

Fig 3.5. A view of grass carp feeding on a weed, Lemna, placed on a platform.

Feeding Groups of Fishes

Fishes are known to use a wide variety of foods which include animal matters, plant matters, decaying organic matters (*i.e.*, detritus) plankton and other micro-organisms living in water. A tabulated information on this aspect for various stages of fish are given below. Some of these subsist on plant matters only while others use animal matters as their food (Alikunhi, 1952; George, 1964). However, a good number of them feed on both animal and plant matters, that is, have taken omnivorous feeding habits. Some fishes use planktons (zoo- and phytoplankton) as their food, but there is lesser number of such types of fishes. A further case of specialisation occurs in few fish called detritus feeders such as *Labeo gonius*. Such fishes depend on decaying organic materials at the bed of column of water, usually intermingled with mud.

Food and Feeding Habits of Some Culturable Carps

Species	Food	Feeding Habits (of adult fish)
(1)	(2)	(3)
Catla Catla	*Fry* : Water fleas, phytoplankton Fingerling : Water fleas, phytoplankton algae, some vegetable debris *Adult* : Crustaceans mainly; also algae, small plants matters, insects, rotifers, vegetable debris.	Omnivorous surface feeder
Labeo rohita	*Fry* : Zooplankton *Fingerling* : Vegetable debris, microscopic plants. *Adult* : Vegetable debris, decaying higher plants, microscopic plants, detritus and mud.	Herbivorous column feeder
Labeo calbasu	*Fingerling* : Same as in rohu *Adult* : Vegetable debris, microso-copic plants, detritus and mud.	Omnivorous detritus feeder
Labeo gonius	*Fingerling* : Algae, vegetable debris, detritus and mud. *Adult* : Vegetable matter, microscopic plants, ditritus and mud.	Herbivorous detritus feeder
Labeo bata	*Fingerling* : As in *L. gonius* *Adult* : As in *L. gonius*	Herbivorous detritus feeder
Labeo fimbriatus	*Fingerling* As in *L. gonius* *Adult* : As in *L. gonius*	Herbivorous detritus feeder
Cirrhinus mrigala	*Fry* : Zooplankton *Fingerling* : Decayed plant and animal matters, algae, detritus, mud. *Adult* : Decayed plant and animal matters, also algae, detritus, organic deposits in mud.	Omnivorous, bottom feeder
Cirrhinus reba	*Fingerling* : Same as in *C. mrigala* *Adult* : Phytoplankton, detritus, organic deposits in mud.	Omnivorous
Hypophthalmi-chthys molitrix	*Fingerling* : Zooplanktons, later phytoplanktons. *Adult* : Blue green algae. Organic detritus, rotifers, copepods, diatoms.	Plankton feeder

(1)	(2)	(3)
Ctenophyryagodon idealla	*Fingerling* : Hydrilla *Adult* : Aquatic vegetation, voraciously feeding on weeds (Hydrilla, Lemna, Azolla, Jassea.	Herbivorous
Cyprinus carpio	*Fingerling* : *Adult* : Algae, Macrovegetation, insects, rotifers, crustaceans.	Omnivorous

Food and Feeding Habits of Some Culturable Non-Carp Fish

Species	Food	Feeding Habits
Clarias batrachus	*Fingerling* : Chironomid larvae, Ostracods, dragonfly nymph, insect eggs and pupae, insect larvae, prawn larvae *Adult* : Insect larvae, shrimps, worms, organic debris, insect eggs.	Carnivorous
Heteropneustes fossilis	*Fingerling* : Fishes, insects, rotifers, crustaceans, algae. *Adult* : Chironomid larvae, Cypris, Copepods, Molluscs, weeds algae	Omnivorous
Channa striaius	*Adult* : Insects, plant materials, crustaecans, worms, insect larvae, fishes.	Carnivorous
Channa marulius	*Adult* : Fishes, crustaceans, insects, worms	Carnivorous
Channa punctatus	*Adult* : Fishes, insect larvae, tadpoles	Carnivorous
Amphipnous cuchia	*Adult* : Predatory on molluscs, worms.	
Wallago attu	*Adult* : Fishes, their fries and fingerlings, insects, tadpoles frogs	Carnivorous
Mystus seenghala	*Fingerling* : — *Adult* : As in *W. attu,* also insects, crustaceans and plant debris.	Carnivorous
Tilapia mossambica	*Fingerling* : — *Adult* : Aquatic plants, filamentous algae. *Fry* : Protozoans, Water-flea.	Herbivorous
Anabas testudineus	*Fingerling* : Animalcules, insect larvae, water fleas *Adult* : Insects, water fleas, vegetable debris, fish, etc.	Omnivorous
Tor tor	*Fingerling* : — *Adult* : Mainly molluscs (40%), also filamentous algae, insects, molluscs, organic deposits of mud	Omnivorous
Tor putitora	*Fingerling* : — *Adult* : Algae, decaying organic matters, insects rotifers.	Omnivorous

In consideration to the type of food used, fishes belong to herbivorous, carnivorous, omnivorous, plankton feeder, and detritus feeder groups. However, from the gut analysis, it appears that very few fishes are exclusively herbivorous or carnivorous types as in most such cases, some percentage of animal matters (in herbivorous type) or plant matters (in carnivorous type) is usually present in the gut (Das and Moitra, 1955; Shafi, 1974).

1. Herbivorous Fishes

These fishes feed mainly on unicellular algal filamentous algae, leaves or twigs of higher aquatic plants and also sometimes planktons. The main food items of herbivorous fishes are the following :

Algae

They are the most common content of the gut of fishes and in some fishes, *e.g.* in *L. calbasu,* they constitute upto 40-50% of the gut content of the fish. The more common species encountered are the following :

Green Algae : Microspora, Oedogonium, Pediastrum, Spirogyra, Ulothrix, Pleurococcus, Tribonema.

Blue-green Algae : Anabaena, Meriomopedia, Microcystis, Nostoc, Oscillatoria, Rivularia, Tetrapedia.

Plant Matters : A good percentage of plant matters are found in the gut content of the herbivorous fishes. However, there may be fluctuation in the percentage composition of plant matters in the food. Usually late monsoon months, such as August-October, have the maximum percentage of these items in India.

Main examples of herbivorous fishes are *Labeo rohita, L. gonius, L. fimbriatus, Ctenopharyngodon idella, Amblypharyngodon mola; Tilopia mossambica, Schizo thorax sinuatus, S. plagiostomus,* etc.

2. Carnivorous Fishes

Food of carnivorous fishes consist of various kinds of animal matters in varying percentage. The main such food items are the following (Fig. 3.6).

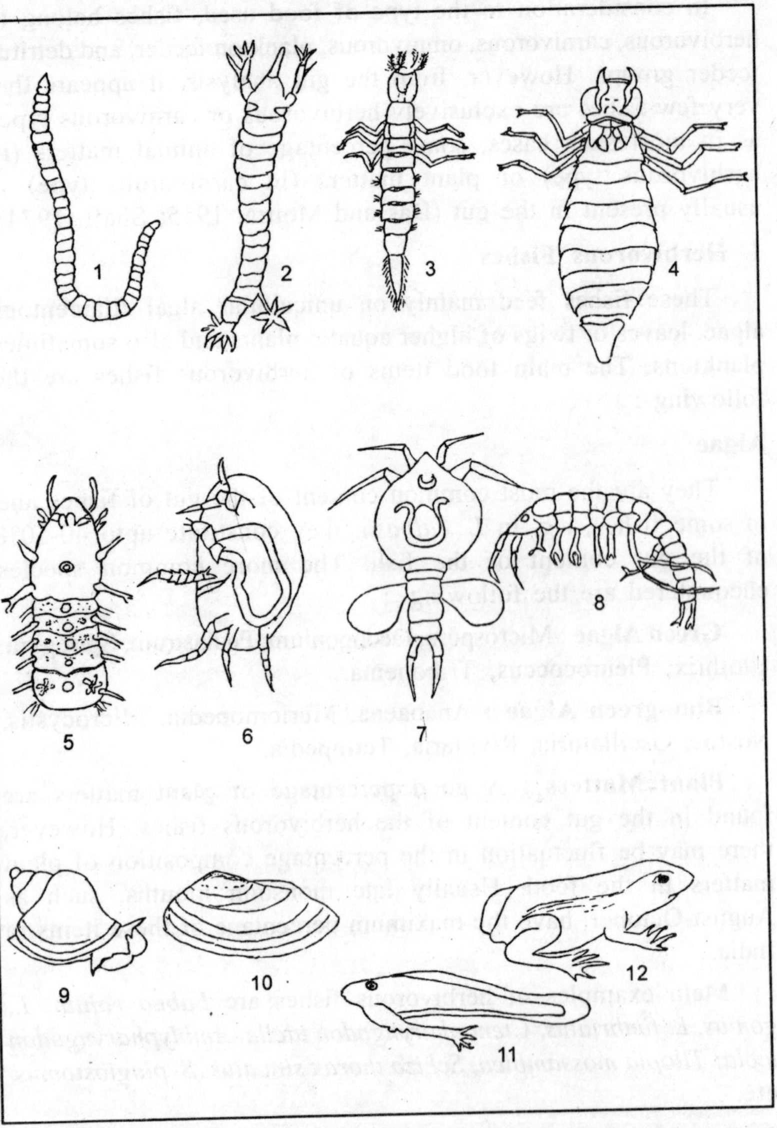

Fig 3.6. Some common animal food organisms : Earthworm (1), mosquito larva (2), Cybister larva (3), Nymph of dragonfly (4), midge (5), Daphnia (6), Cyclop (7), Amphipod (8), Pond snail, Limnaea (9), Unio (10), Tadpole (11), Frog (12).

Crustaceans. Ostracods, Copepods (*e.g.* Cyclops, Diaptomus, Pseudodiaptomus), Daphnids, Cladocera, Alona, Chydorus, Sidda, Daphnia, Bosmina.

Insects. Notonectids, mosquito larvae, *Dytiscus cybister,* Elmis, Gyrinus, Hemispera, (*i.e.*, Ranatra, Corixa, Notonecta). Diptera (Chironemid larvae mosquito larvae, adult form of Chironomus, Culex, Eristalus, Simulium), Odonata (nymph of dragon fly, Aeschna, Cordulum, Sympetrum, Mesothemis).

Rotifers. Keratella, Brachionus, Asplanchna, Notholca, Distyla, Polyrathra, Euchanis, Gastropus.

Annelids. Earthworms, Tubifex

Molluscs. Bivalves (freshwater mussel), snails (Limnaea, Planorbis, Physa, Vivipara, Pila).

Fishes. Larval forms (fries and fingerlings), small fishes of all kinds.

Amphibians. Tadpole larvae, small frogs.

Surface feeding carnivore may be recognized by the presence of surface fish and animals in its gut contents, a mid-feeding carnivore by their column feeding fishes, the presence of surface to bottom food animals in gut contents, and the complete absence of sand. A bottom carnivore can always be identified by the presence of bottom fauna as well as the presence of sand in the gut contents.

According to these identification criteria, three groups of carnivore fishes are exemplified :

(i) Carnivore — bottom feeder, viz., *Callichrous pabda.*

(ii) Carnivore mid-feeder, viz., *Wallago attu, Mystus cavassius* and *M. Viltatus.*

(iii) Carnivore bottom feeder, viz., *Notopterus chitala, Channa marulius, C. striatus, Clarias batrachus.*

As per niche, the freshwater fishes occupy in water for getting food, the fishes have been divided by fishery scientists into three distinct groups.

(1) **The Surface feeders.** Such as *Catla catla, Hypophthalamichthys molitrix, Oxygaster bacaila, Puntius ticto, Chanda*

nama, C. ranga, Glossogobius giuris, Gadusia chapra, Hilsa ilisha, etc.

(2) The column feeders, such as *Labeo rohita, Puntius sophore, Tor tor, Wallago attu, Mystus cavassius* and *M. vittatus.*

In this category, some fishes are neither true surface feeders, nor true bottom feeders and mostly depend on the organisms of the midwater. Such species are also called midwater feeders.

(3) The bottom feeders collect their food from the bottom area of ponds. Fishes in this category are *Labeo calbasu, L. bata, L. gonius, Cirrhinus mrigala, C. reba, Puntius sarana, Amblypharyngodon mola, Clarias batrachus, Channa marulius, C. striatus, Heteropneustes fossilis.*

This is an important fact that the surface feeders are either omnivorous or carnivorous, whereas the midwater, and bottom feeders may be herbivorous, omnivorous or carnivorous.

REFERENCES

Alikunhi, K.H. 1952. On the food of young carp fry J. zool. Soc. India 4 : 77-84.

Bhimachar, B.S. 1971. Food of the Carp hatchlings in fish culture. Indian Fing. 20 (12) : 45-46.

Chakrabarty, R.D., Sen, P.R., Chatterjee, D.K., and Kowtal, G.V. 1973. Observations on the relative usefulness of different feed for carp spawn and fry. J. inland Fish Soc. Ind. 5 : 182-88.

Bhowmick, M.L. 1968. Environmental factors affecting fish food in freshwater fisheries, Kalyani, West Bengal. Unpublished thesis. Univ. of Kalyani : 238.

Das, S.M. and Moitra, S.K. 1955. Studies on the food of some common fishes of Uttar Pradesh, India. 1. the surface feeders, the mid-feeders and the bottom feeders. Proc. Nat. Acad. Sci. India 25 B (1 and 2) : 1-6.

George, Michael, R. 1964. Limnological investigations on pond plankton, macrofauna and chemical constituents of water and their bearing on fish production. Unpublished D.Phil. Thesis, University of Calcutta.

——. 1969. Seasonal trends in physico-chemical factors and plankton of freshwater fish pond and their role in fish culture. Hydrobiologia 33 (1) : 145-60.

Shafi, Md. 1974. Histology and histochemistry of the digestive organs of some freshwater teleosts. Ph.D. Thesis, Aligarh Muslim University, Aligarh (India).

Sinha, V.R.P. and Gupta, M.V. 1975. On the growth of grass carp *Ctenopharyngodon idella* Val. in composite fish culture at Kalyani, West Bengal (India). Aquaculture 5 : 283-90.

——. Gupta, M.V., Banerjee, M.K. and Dhirendra Kumar. 1973. Composite fish culture at Kalyani, West Bengal. J. Inland Soc. India 5 : 201-07.

Srivastava, V.K. 1956. Bottom organisms of a freshwater fish tank. Curr. Sci. 23 : 158-59.

——. 1959. Studies of freshwater bottom fauna II. Qualitative composition and variation of the available food supply of fishes. Proc. nat. Acad. Sci. Ind. 29 : 207-16.

4

Natural Breeding

The annual quantum of the fish production as well as average fish consumption rate in India are conspicuously low as compared to the figures available from Japan, China, Indonesia and U.S.A. In consideration to the ever-increasing rise of the population figures, it is estimated that by 2000 A.D., the requirement of fish could reach 13.0 million tonnes. The stipulated requirement can be fulfilled by advanced planning and taking necessary measures for the development of inland fisheries. It is an admitted fact that there was no scope of culture practices in sea water as the fishing operation areas have no boundaries and seed collection and rearing of the spawn or fry is not possible. Obviously, the whole thrust is on inland water sources for procurement of fish seeds.

For achieving target of sufficiently high fish production figure by 2000 A.D. the foremost requirement is the availability of sufficient quantity of fish seeds, and the same must be of high quality, having less mortality chance. Figure of fish seeds through natural sources and by induced breeding or hypophysation technology was more than three decades before, inadequate, being 5,505 million of spawn, 1,651.8 million of fry and 8,250 million of fingerlings (Ref. National Fish Seed Committee, 1966). However, the recent figure (1996-1997, as target) stands for 16,500 million. This increase in production figure was attained by adopting the improved method of fish breeding. (Choudhary, 1960 Alikunhi *et al.*, 1963) and application of latest hatching technologies (Sinha and Ramachandran, 1985) to have high quality of viable fish seeds for stocking ponds. The account endorsed here includes valuable informations from various sources.

Favourable Factors for Natural Spawning

Several factors are directly or indirectly concerned for influencing maturity and spawning of fish. Among such factors, the main are rain (*e.g.* monsoon season), flood, light (its fluctuation level), temperature (in optimum maximum-minimum thresholds), dissolved oxygen concentration, pH, turbidity, water current, dissolved salts, presence of shallow areas and, nevertheless, the repression factor (which is specific, being excreted into the water by the potentially breeding species of fishes). Out of these factors, the rain, flood and temperature are prime factors to influence the fish directly to indulge in breeding actions. This begins first by developing symptoms of maturity, then there is reflection of some signs in posture and behaviour and finally spawning, takes place in a choiced place.

Internal Factors

Among the internal factors, most concerned factor is the endocrine secretion, as this factor is most highly responsible for the attainment of maturity and spawning action in fishes. The sex-stimulating harmones of the pituitary gland play a pivotal role in the development and maturation of gonads and spawning in fishes. However, the endocrine secretion is also under the influence of certain external factors like rain, light, temperature etc.

The carp culture practice is popular throughout the country. A regular and dependable source of quality fish seed is by inducing the fish to breed in ponds. This inducement is either by bundh breeding or by adopting the technique of pituitary injection (*i.e.* by hypophysation). The age and size of the carps vary according to the climatic and ecological conditions of the habitats, the availability and utilization of food, and also as per inherent genetical characteristics as mentioned in Table 4.1.

Breeding Season

The breeding season for all major carps (excluding common carp) in India is monsoon season (July-August), but the breeding season may begin a few months earlier, or it may prolong as per monsoon extension, such as observed in Assam areas (India). All

Table 4.1

Fish	Maturity time-period (years)	Specific features
Catla	2	Males mature earlier (within 1 year)
Rahu	2	Males may mature even in 1 year
Mrigal	2	Both sexes mature in 2 years
Silver carp	1-2 In China 2-6 In USSR 3-10	Males mature 1 year before
Grass carp	2	Males may mature in 1 year
Common carp	1-2	Males mature within a year

above mentioned species breed naturally in flooded river conditions during monsoon, but the common carp breeds in ponds, having two breeding season and monsoon and post-winter periods (February-March). *Catla catla* has been reported sometimes to spawn twice a year in Penninsular India, *i.e.,* during the south-west and north-east monsoon.

Natural Breeding in Rivers

It has been observed that all major carps have almost the same breeding habits. With the onset of monsoon, the rivers have raised level of freshwater and a current is produced due to flooded condition. The breeding carps get induced to ascend against the current of water and move to an adjacent inundated shallow areas of the river for the purpose of spawning (Fig. 4.1). In this favoured conditions, the mature males and females indulge in sex play which probably stimulate the pituitary gland to secrete hormones and thus incurred inducement leads to ovulation and spawning (Khanna, 1957; Sinha *et al.,* 1974).

The period of peak maturity of mrigal and gonius is May and that of rohu is June. In South Bihar (now Jharkhand State), the spawning of carps extends from April to August, in Orissa, Uttar Pradesh, Madhya Pradesh, the breeding season is from June to

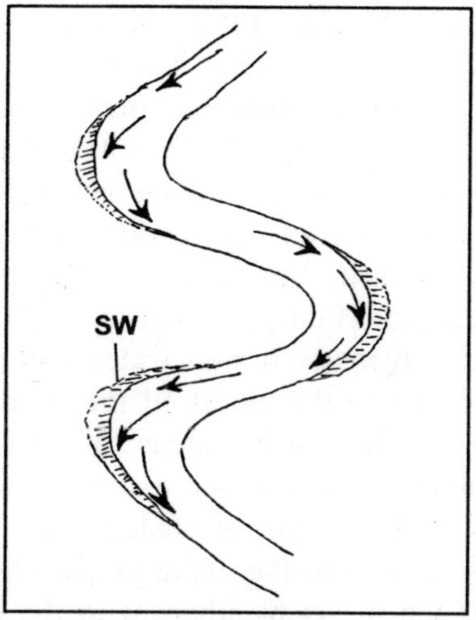

Fig. 4.1. Natural Breeding Site : River bank with shallow water spread area (SW) during monsoon.

August and in Andhra Pradesh, Mysore and Punjab, the monsoon begins later and the breeding season lasts from July to September.

Little information is available on the hydrological conditions of the riverine spawning grounds of major carps. The chemical composition of a particular spawning ground in West Bengal was found to be like this pH : 7.6; Temperature : 30°C; Oxygen (dissolved) : 6 ppm; and CO_2 (free) : 2.1 ppm (Sinha *et al.*, 1974).

BREEDING OF CARPS IN BUNDHS

The major carps breed also in special type of perennial and seasonal ponds or impoundments, locally known as bundhs in West Bengal, Bihar and Madhya Pradesh. Functionally they are of two kinds : (1) 'perennial bundhs or wet bundhs' and (ii) 'seasonal bundhs or dry bundhs'. These ponds or bundhs receive rain water with washings from the widely spread catchment areas during rainy seasons, particularly after heavy shower. The large shallow marginal water areas of these ponds serve as spawning grounds for the major carps. As riverine conditions are stimulated during the monsoon season owing to accumulation of rain water, these sites are preferred by fishes for breeding.

Wet Bundhs. Wet bundhs are perennial ponds located in the slope of vast catchment area. They have proper embankments, are provided with an inlet towards the uplands and with an outlet on the opposite site (Fig. 4.2). During heavy shower, a great portion of the bundh becomes inundated and the excess of water flows out through the outlet. However, during summer, the extensive shallow area dries up, but the main pond retains water throughout the year. In Midnapore district of West Bengal the streamlets, through which fresh rain water with washings from the upland are rushed into the mouth of the bundhs is known as 'dhal', whereas the shallow water areas of the bundh, is called 'moan' and the outlet through which excess of water is drained is known as 'bulan'. The moan or shallow area of the pond is the main spawning ground of the major carps. The moan may be located either near the outlet or adjacent to the inlet. A sort of protection is provided to the outlet by a bamboo fencing called 'chhera'. The flow of water through the outlet can be controlled

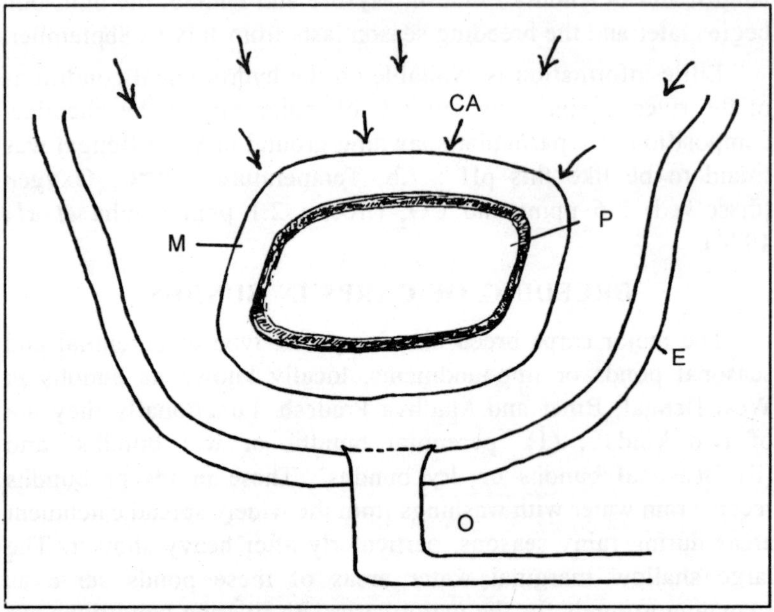

Fig. 4.2. Another Natural Breeding Site : A wet bundh with catchment area (CA), Embankment (E), pond (P), Moan (M), water outlet (O).

by plugging the 'chhera' with straw and mud. A wet bundh can be of any shape and dimension. Normally a catchment of 20th to 100 times the area of bundh is considered ideal for breeding (Singh *et al.,* 1979). Examples of good bundhs are also large irrigation tanks having rich coverage of fresh plants or fields in the catchment areas. These may also serve as ideal wet bundhs (Kamal *et al.,* 1979).

Dry Bundhs. Dry bundhs is shallow seasonal pond, much smaller than the wet bundh, bounded on the three sides by embankments which impound fresh water from the catchment area during the monsoon season. As it remains dry for the greater part of the year, it is known as 'dry bundh' (Fig. 4.3A,B). It has been observed that in West Bengal, if the catchment area is more than five times the main area of the dry bundh, it is considered least suited for breeding, while in Madhya Pradesh the ideal ratio between these is 25:1. Related to this fact, the dry bundh area in Madhya Pradesh is much bigger, being 0.2-2.5 hectares whereas in West Bengal, the area is 0.1-0.5 hectare. However, several modifications can be found in the typical wet and dry bundhs and some variations in the method of breeding carps also exist (Dubey, 1969). The dry bundh is preferred over the wet bundh in consideration to purity of seeds for the reason that in the deep wet bundhs, many varieties of unwanted fish live along with major carps so that after breeding, the seeds get mixed up. Moreover, the wet bundhs, being larger and deeper, are not operational easily for getting seeds.

Breeding in Dry Bundh in West Bengal

The bundh are of varied size, having catchment areas of about five times the main bundh area. They are repaired before the onset of monsoon, the inlets and outlets are given proper shape, and provided with a bamboo or velon screen so as to obstruct the escape of breeders and the entry of unwanted species. After accumulation of sufficient water in the bundhs, brooders from the nearby ponds are transferred. The congenial environment of the bundh induces the male and female brooders to indulge in sex play for sometime and then breed which generally happens during the early hours of the morning (Ganpati and Chacko,

1954; Singh *et al.*, 1979). This may continue throughout the day and in all positions of the bundh. The released eggs are collected by mosquito nylon net or velon net and transferred to earthen pits especially dug in earth in an adjacent place. These pits act as hatcheries. The hatcheries generally have shallow water with silt particles in water. Therefore, when rain is over and in the clear weather the temperature is high, the eggs and spawn may have adverse effect of exposure to high temperature. As such, mortality may occur.

Breeding in Dry Bundh in Madhya Pradesh

In Madhya Pradesh, bundhs have water surface area of 0.2 to 2.50 hectare. In most cases, bundhs are provided with sluice gates in the deeper region for dewatering the bundh flow after each spawning episode and this provision is also helpful in removing the spent fish and collecting the fry and fingerling, if living in this bundh. During rain, the bundh gets high accumulation of water. There is provision of one or two surplus weirs of variable size (in 15-30 cm size) to allow surplus water to flow out, when need arises. The provision of iron grating frames at waste weirs, covered with small-meshed nets or mosquito netting for use for few days during spawning period, helps to check the escape of eggs and spawn with the outgoing flow of water. To facilitate stocking of the breeders temporarily, prior to transfer in the bundhs, a few cemented cisterns or earthen ponds are constructed and these are called storage tanks. These can also be used to store spawn or fry prior to their transfer to fish farm. These and some other aspects are suitably displayed in figures (Fig. 4.3A, B).

The breeders (males and females) collected from the reservoirs or wet bundhs are stocked separately in the storage tanks. The feeds in the forms of oil cake at the rate of 3-5 of their body weight is given. On accumulation of rain water in the dry bundhs, reaching 1 to 1.5′ depth, selected number of breeders, 1 female and 2 males, are introduced and a close watch is maintained, from a specially built observation tower regarding breeding actions, *i.e.,* scene of spawning by females. Eggs or spawn are collected with the help of nylon net (mosquito net) and transferred to the double walled hatching hapas, as described

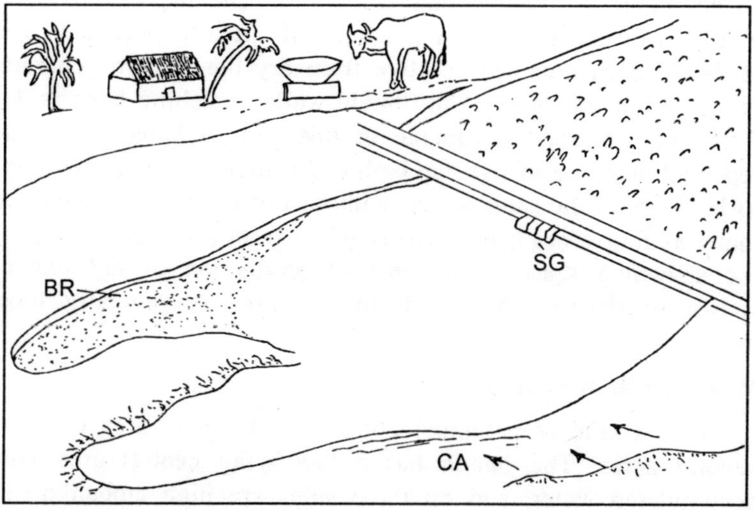

Fig. 4.3A. Another Natural Breeding Site : A dry bundh with catchment area (CA), Breeding ground (BR) and sluice gate at bundh (SG).

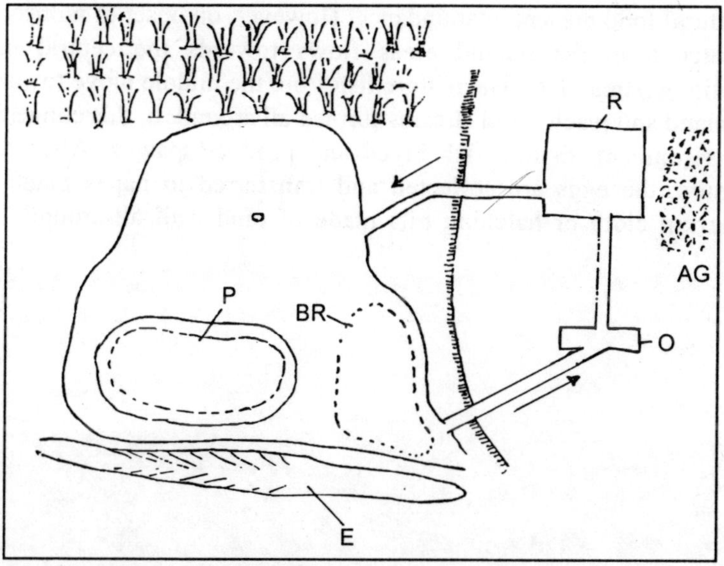

Fig. 4.3B. Yet another site : A dry bundh of West Bengal with the main pond (P), Breeding ground (BR), Embankment (E), Water Reservoir (R), outlet (O), Agriculture land (AG).

ahead about its construction (Fig. 5.5). The hapa is fixed in bundh itself or in cement hatcheria if already provided. The fertilized eggs obtained as alive from dry bundhs are generally 60-68%, but the survival in double walled hatching hapa is 32-50% only. A great advantage of this sort of breeding is that repeated breeding of fish can be brought in action with fresh sets of breeders in the same water, without waiting for rain. Thus, as many as four crops can be easily taken from a dry bundh during the spawning season. The cost of production of millions of spawn of desired species of major carps is generally most moderate.

Wet Bundh Breeding

It is a kind of perennial pond or tank, in an area to cover several acres. The bundh has a low lying central area with accumulated water, and on three sides are high embankments (Fig. 4.2). In summer, a part of the upland area dries up and the farmers use it for cultivation, but in the low lying middle portion some water is always present where mature fish enjoy life with natural food present in abundance. However, on onset of monsoon, water from the upland areas flows towards the central part forming streamlets. The mature fish from the middle move towards upland and reach small streams formed all around. In this congenial environment, mature fish breed and produce spawns. After few hours, the eggs are collected and transferred to hapas made of markin cloth or hatching pits made of mud wall all-around.

5

Induced Breeding

Management of fresh water ponds involves many requirements such as pond preparation, rational application of fertilizers, control or eradication of unwanted aquatic vegetation, pests and predator organisms etc. However, stocking with good quality of seed (*e.g.* fast growing fish seed) at appropriate rate and ratios and post-stocking care of the same are the foremost requirements in view of the fact that the success of fish culture has to be evaluated from the quantum of fish field. It needs cognizance that all efforts on intensive culture practices can get defeated if the quality of fish seeds, to be stocked in nursery ponds, is not of high grade. The seeds collected from the rivers are generally of mixed variety, containing also seeds of predaceous and trash fishes. The best source of pure quality of fish seed is through induced breeding technique. Induced breeding by pituitary extract injections, also referred to as hypophysation, is an effective way of attaining pure seed of cultivable carps. The hypophysation of fishes provides an excellent opportunity of the successful application of a laboratory technique to the solution of a major practical problem of production of quality fish seed.

Another point of consideration is that most Indian major carps, which represent important culturable food fishes of India, are not able to breed in ponds. This factor is also largely responsible for the acute shortage of fish seeds in the country. These two considerations and also growing urge among fishery scientists led to evolving, for the culture practices, the modern techniques of pituitary treatment commonly called hypophysation, or simply induced spawning by pituitary hormone injection (Alikunhi *et al.*, 1960; Chaudhuri, 1976). Induced spawning by pituitary hormone injection has paved the way for commercial production of pure fish seed on a large scale. The technique has been most efficiently applied by the fish farmers of West Bengal (specially of Midnapore and Bankura districts), so much so that they have brought fame to their state and achieved highest rank as fish seed producer in the country, having a state share of about 70% of the total production of fish seeds.

General Account

In India, first attempt was made as early as in 1937 in this direction but a real and first success in this direction was achieved only in 1957 at the Central Inland Fisheries Research Station, Calcutta. Since then, further efforts were made by research scientists to improve the technique and get a desired result. Success to a great extent has been achieved now, and there is a network of centres for the production of fish seeds of high quality by adopting this induced spawning method.

Process of Induced Breeding. Generally mixed culture of selected non-predaceous and fast growing species of fishes with short food cycle is prepared, as this gives high yield. The Indian major carps, catla (*Catla catla*), rohu (*Labeo rohita*), mrigal (*Cirrhinus mrigala*) and exotic carps such as common carp (*Cyprinus carpio*), silver carp (*Hypophthalmichthys molitrix*) and grass carp (*Ctenopharyngodon idella*) qualify these requirements and are well-known to give high yield.

Collection and Preservation of Pituitary Gland

Pituitary gland extract is the most commonly used inducing agent for the Indian major carps. Pituitary glands are collected from the ripe fish either belonging to the same species or a closely related one (Fig. 5.1). The glands from the major carps is collected just prior to or during the breeding season. As the common carp (*Cyprinus carpio)* breeds throughout the year, its mature individuals can be available whole year so that the glands can be collected from them. Pituitary glands are collected from freshly killed mature carps, or from induced bred fishes immediately after spawning; however, those, collected from ice preserved specimens are also useful. Immediately after removal, the glands are preserved in absolute alcohol or acetone.

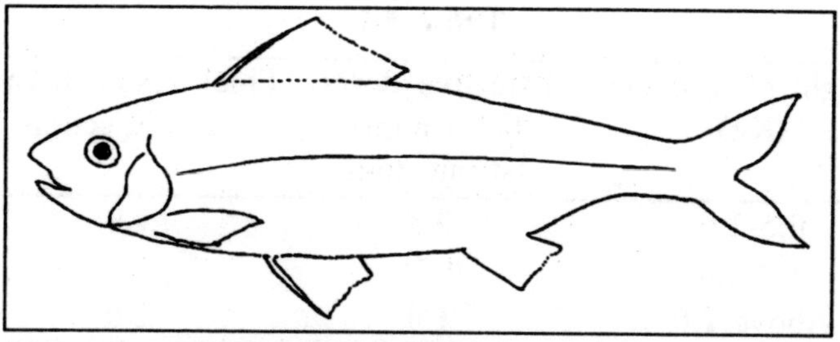

Fig. 5.1. Outline view of a ripe carp.

The ideal method of gland collection is by chopping off the scalp of the fish skull right above the eye by inflicting an oblique stroke of a sharp knife. After removing the scalp, the grey matter and fatty substances lying near the brain are cleaned with the help of a piece of soft cotton. Thus the brain gets well exposed. It is now lifted up by detaching it from the nerves. The gland with the covering membrane is then picked up and gently removed.

The alternative method of collection of pituitary gland is from the posterior lobe of the skull, known as foramen of magnum. In this method, the fish is beheaded first. As the brain lies in the ventral side of the foramen, the grey matter and fatty substances are first removed by manipulation with the forceps, and, then, the brain is lifted above carefully. The gland lying at the lower (*i.e.* ventral) side of the brain is then picked by using a pair of fine tweezers (Fig. 5.2).

Preservation of Pituitary gland. Soon after collection, the pituitary glands are preserved in absolute alcohol for defattening action. After changing the alcohol, 2-3 times (consuming about two hours time in total) the defatted and dehydrated glands are stored in small dark vials and kept in refrigeration at low temperature or in desiccator at room temperature. Now the doses are amputed according to their proper weight given in the ampule.

Preparation of Pituitary extract. Preserved gland after removal from the alcohol is weighed in a chemical balance and required numbers are then macerated in a tissue homogeniser by adding a measured quantity of distilled water. If homogeniser is not available, the purpose of homogenisation of the gland can be accomplished by using an ordinary mortar and pestle. The requisite for dilution of pituitary gland is as given in Table 5.1.

Table 5.1

Weight of spawner (Kg)	Effective dose/ fish (in ml) Ist injection	Final dose/fish (ml) 2nd injection
0.5-1.0	0.5	1.0
1.0-2.0	0.75	1.5
Above 2.0	1.0	2.0

After homogenisation, the solution is centrifuged and the supernatant fluid are drawn in hypodermic syringe for injection.

Method of Breeding

The breeders (individuals selected for hypophysation) are taken out of the hapa and a desired quantity of the pituitary extract is injected intramuscularly. Immediately afterwards, the recipient fish is retained to the hapa. It is important to consider that the pituitary extract is not injected to induce gonadal development, rather, only to initiate ovulation in females and spermeation in males. The gland of the same species of the carps (homoplastic) or different species (heteroplastic) are used for administration of hormone. Expert's opinion is that a cool rainy day, when water temperature is below 31.5°C, is ideal for the induced breeding practice.

Selection of breeders for hypophysation. For having proper quality of fish seeds, the breeder individual should be judiciously selected for hypophysation. Better results can be achieved by stocking medium and large sized fishes, 4-5 months before the onset of monsoon. An ideal stock manipulation in brood fish pond is given in Table 5.2.

The pond should be fertilized with organic fertilizer periodically. Breeders should be artificially fed with oil cakes and rice or wheat bran in the proportion of 1:1 in the amount of 1 — .5% body weight to maintain them in ideal condition. The condition of the fish may be examined periodically by netting action. Fully ripe females having a round, soft bulge, elastic abdomen and slightly swollen pinkish vent (anal opening), and males with freely oozing milt on gentle pressure are selected for administration of hormone, *i.e.*, for hypophysation. A good male breeder can

Table 5.2
Stock Manipulation in a Brood Fish Pond

Primary brood Fish	Stock density kg/ha	Secondary brood fish	Stock density kg/ha
Catla	500 to 1000	Grass carp Silver carp Rohu	1000 to 1500
Silver Carp	750 to 1300	Grass carp Rohu Mrigal Common carp	700 to 1250
Grass carp	1000 to 1500	— — — —	500 to 1000
Rohu	Do not require	Any manipulation	
Mrigal	Do not require	Any manipulation	

also be distinguished by feeling the denticulation on the dorsal side of the pectoral fins (Fig. 5.2A, B).

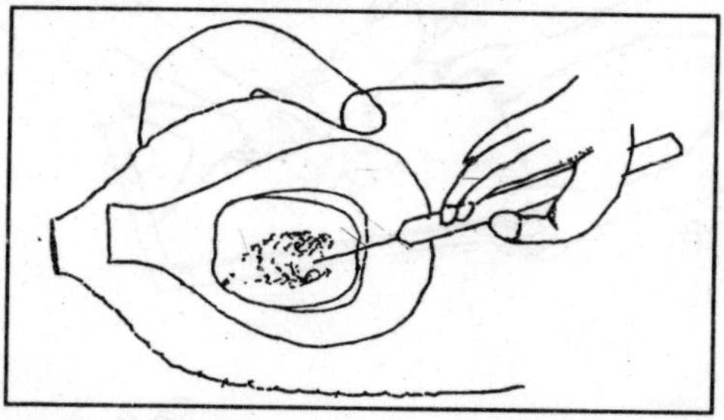

Fig. 5.2A. Dissection of the head region for exposure of the pituitary gland.

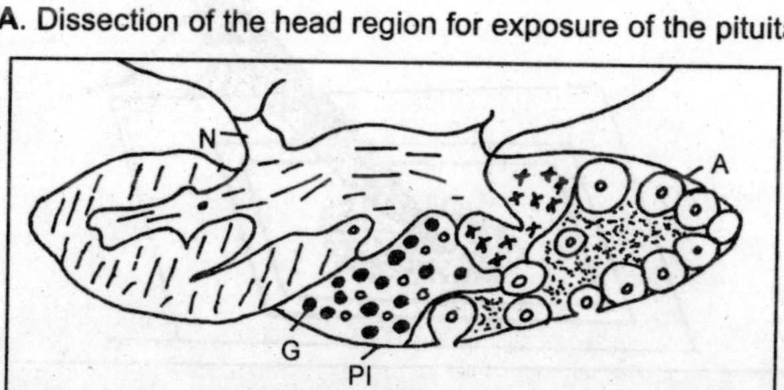

Fig. 5.2B. A view of histological structures of the pituitary gland of a carp, Neurohypophysis (N), Gonadotrophs (G), Adenohypophysis (A) and Pars Intermedia (PI).

As an ideal practice, one female and 1-2 males (3-4 in case of silver carp) are introduced in a breeding hapa or chinese hatcheria with provision of flowing water. Breeders can also be kept in cemented cistern or plastic pool but these choices are not preferable.

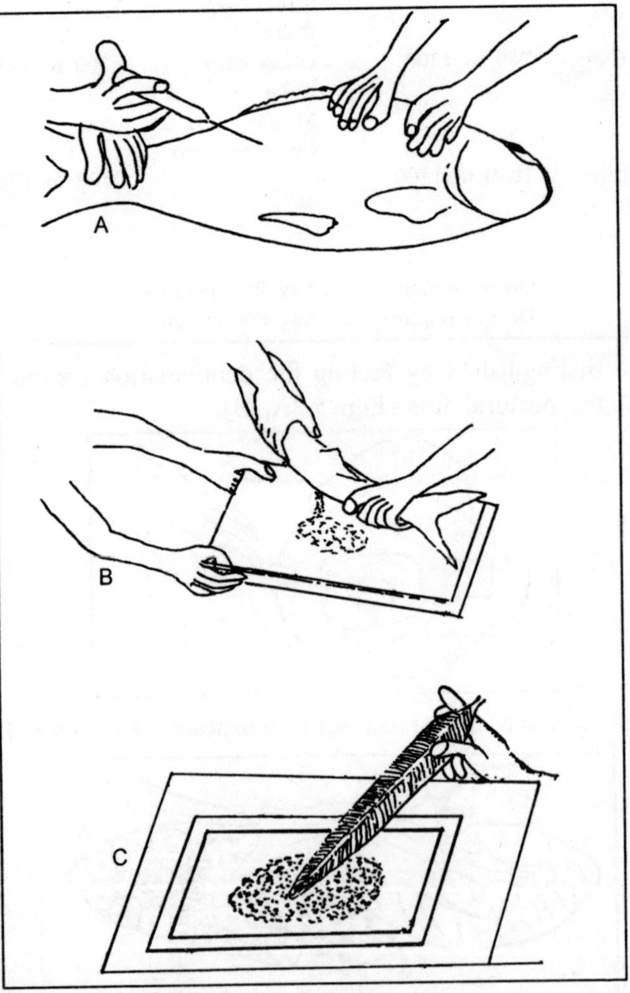

Fig. 5.3A. A view of pituitary extract injection to a carp (A), the release of eggs (B), mixing of eggs and milt (C).

Administration of hormone. The usual practice for hormone injection to major carps is that a preliminary dose of 1-3 mg gland/1 kg body weight, and, following it 4-6 hours afterwards, a second dose of 5-12 mg/kg weight are administered to the female first (Fig. 5.3). The first dose is called primer and the second dose is booster. The males on the other hand, receive a single dose of 1-3 mg gland/1 kg body weight at the time of application of second injection to the female.

In the case of chinese carps, a total of 9-14 mg gland/kg weight is administered to the female in the stage, 3-4, preferably @ 6-10 mg gland/1 kg body weight. Interval between two injections are 9-6 hours. However, the male chinese carp receives single dose of 2-6 mg/gland/kg weight at the time of first injection to the female. Doses are decided on the maturity condition of fishes as well as prevalent weather.

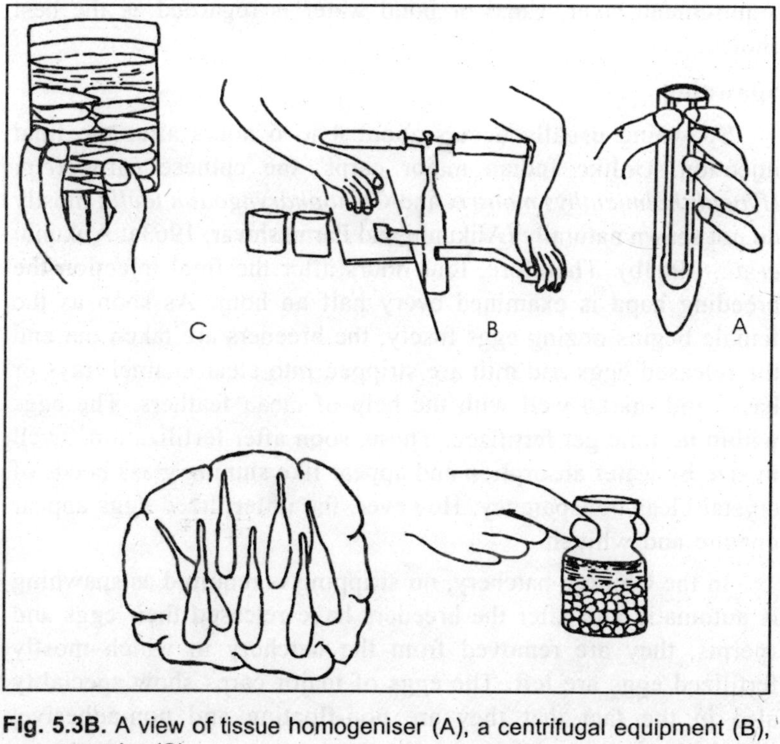

Fig. 5.3B. A view of tissue homogeniser (A), a centrifugal equipment (B), ampoules (C).

Specification of Breeding Hapa

After the administration of pituitary extract, the breeders are kept inside a breeding hapa, as mentioned before also. A breeding hapa is made of thin markin or mattie cloth the size of which varies from 2 m × 1 m × 1 m to 4 m × 2 m × 1 m (Fig. 5.4). The choice for size of the hapa is kept related to the size of the brood fish. The hapa remains firmly stitched on all sides except at the top for introducing the breeder fish inside the hapa. Usually one set of breeders is released in each breeding hapa and soon after release, the opening of the hapa is tied well so that the breeders cannot escape.

An alteration to breeding hapa of cloth make is the cemented cisterns or plastic pools in size as big as breeding hapa. The water column in the hapa should be maintained at 30-35 cm, preferably with provision of slight circulation of flow of water. To meet this requirement, river, canal or pond water is regarded as the best choice.

Spawning

Spawning usually occurs about 4 to 6 hours after the final injection. Unlike Indian major carps, the chinese carps (*e.g. Hypophthalmichthys molitrix* and *Ctenopharyngodon idella* mostly do not spawn naturally (Alikunhi and Parmeshwar, 1963a; Alikunhi *et al.,* 1963b). Therefore, four hours after the final injection the breeding hapa is examined every half an hour. As soon as the female begins oozing eggs fusely, the breeders are taken out and the released eggs and milt are stripped into clear enamel trays or basin and mixed well with the help of clean feathers. The eggs within no time get fertilized. These, soon after fertilization, swell in size by water absorption and appear like shining glass beeds of crystal clear transparency. However, the unfertilized eggs appear opaque and whitish.

In the Chinese hatchery, no stripping is required as spawning is automatic and after the breeders have released their eggs and sperms, they are removed from the hatchery in which mostly fertilized eggs are left. The eggs of major carps show speciality also in the fact that they are non-floating and non-adhesive, excepting the case of common carp.

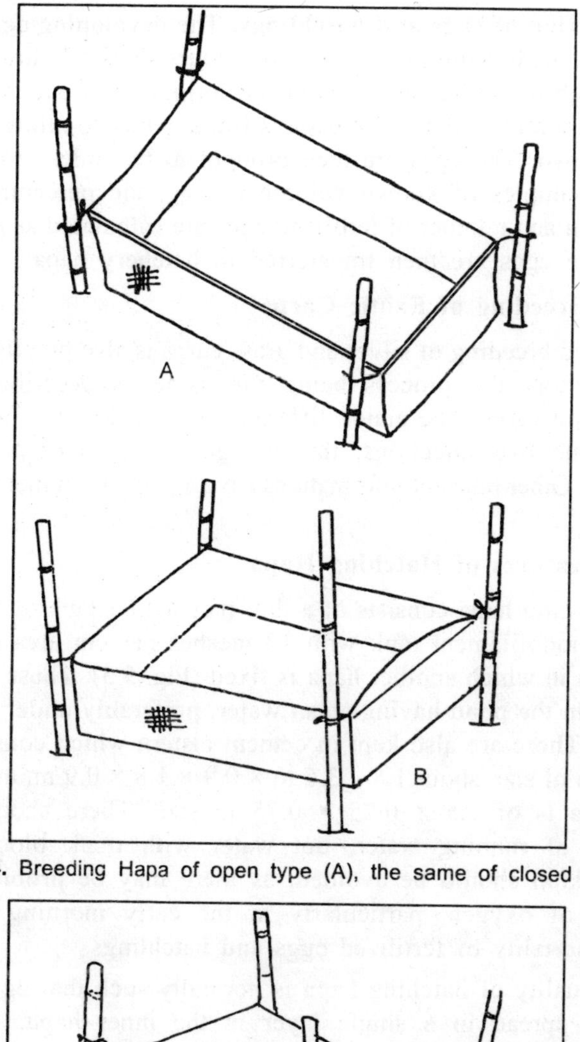

Fig. 5.4. Breeding Hapa of open type (A), the same of closed type (B).

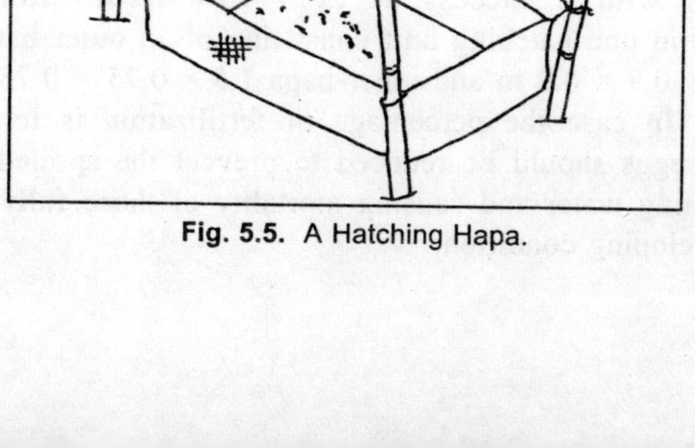

Fig. 5.5. A Hatching Hapa.

Collection of eggs and hatchlings. The developing eggs are retained in the breeding hapa for 4 to 6 hours so as to allow them time enough to get hardened. After the formation of embryo, the brooders are removed and the hapa is lifted gently to allow water to drop down. The eggs are then brought to the side. From the count of samples of known volume of egg, the percentage of fertilization and number of fertilized eggs are calculated as per ml figure. The eggs are then transferred to hatchery hapa.

Induced Breeding of Exotic Carps

Induced breeding of silver and grass carps is also practised by hypophysation, the process being the same as described for Indian major carps. The major difference lies in the time interval between the two injections, this being 9-6 hours (instead of 6-4 hours). Other relevant informations have already been mentioned before.

Special Features of Hatching Hapa

A hatching hapa consists of a thin rectangular tank of cotton cloth or monofilament tank with 12 meshes per cm area called outer hapa in which another hapa is fixed (Fig. 5.5). These hapas are fixed in the pond having clean water, preferably under shade of a tree. These are also kept in cement cistern which consist of outer hapa of size about 1.8 – 3.6 m × 0.9 × 1.8 × 0.9 m, and the inner hapa is of 1.5 × 0.75 × 0.75 m size. There should be provision of running water, but water with thick bloom of phytoplankton should be avoided, as there may be problem of depletion of oxygen, particularly in the early morning hour, causing mortality of fertilized eggs and hatchlings.

The quality of hatching hapa is normally such that eggs are uniformly spread in a single layer in the inner hapa. About 50,000 – 1,00,000 fry can be introduced in one hatching unit. Eggs with a success of even 90% fertilization can be hatched in one hatching unit consisting of an outer hatching hapa, 1.5 × 0.9 × 0.9 m and inner hapa 1.5 × 0.75 × 0.75 m in dimension. In case the percentage of fertilization is less, the density of eggs should be reduced to prevent the spoiled eggs from polluting water and causing mortality of those fully alive and in developing condition.

It has been found that most of Indian major carps hatch out in about 14-18 hours after fertilization, but those of chinese carp consume more time, *i.e.,* usually 10-20 hours after fertilization to hatch out. The exact time period depends on water temperature. These tiny organisms pass through the meshes of inner hapa into the outer hapa. When all the eggs have hatched out, the inner hapa with dead eggs and egg shell is removed. Hatching takes place in the outer hapa for about 2-2½ days. During this period, the yolk is absorbed, the mouth of the pectoral fin is formed and the young one begins active movement in water (Bhowmick, 1976; Sinha and Ramachandran, 1985).

Induced Breeding of Common Carps

Out of these varieties of common carps, the scale carp (*Cyprinus carpio var communis*) is extensively cultured in the plains of India, usually in combination with the Indian major carps, this carp breeds naturally in its natural habitat, *i.e.*, ponds and tanks as mentioned before. Under controlled conditions, the common carp is successfully induced to breed by simulating its natural spawning habitat in a relatively small confined space, without going for injecting pituitary extract dose (Fig. 5.6). Lakshmanan *et al.* (1971), Sinha *et al.* (1974) and Bhowmick (1976) have provided useful informations in this respect.

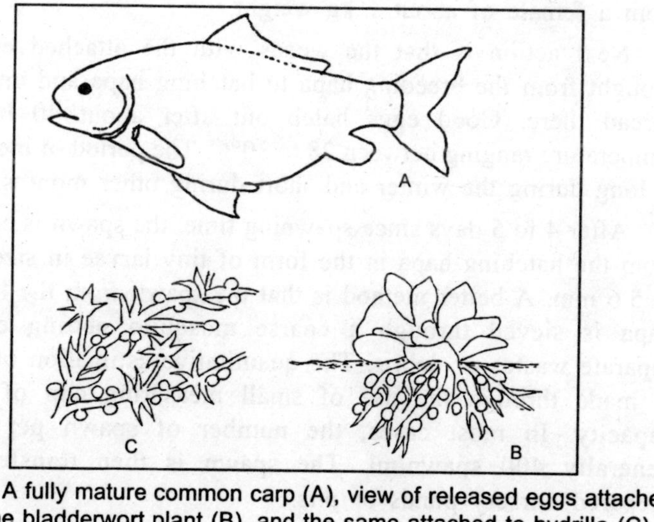

Fig. 5.6. A fully mature common carp (A), view of released eggs attached to the bladderwort plant (B), and the same attached to hydrilla (C).

The common carp is bred in hapa, which is fixed in the pond. For a successful spawning, males and females in equal proportion by weight are released inside the hapa. As the eggs are sticky, soon after the introduction of breeders, the requisite quantity of weeds, *e.g.* water hyacinth or hydrilla is placed inside the hapa with the effect that the eggs stick on the twigs or branches and roots. The cover of the hapa is then tied closely to check the possibility of jumping out of the breeder. The expert advice is to release brood fish in the hapa in the evening time. It has been observed that the fish generally spawn at intervals (not once in all) within 6-10 hours after being released in the hapa. The released eggs are 1.0 to 2.0 mm in diameter and they respond as sticky to the weeds, as shown in figure (Fig. 5.6B,C). The fertilized eggs of good quality are dirty pale yellow in colour, whereas the unfertilized eggs look opaque white in colour, being more prominently visible than the fertilized ones. The common carp, as said before about the Indian major carps, can also be bred in plastic pools or cistern made of cement.

After the spawning action is over (*i.e.,* after 6-10 hours) the brood fish of common carp are removed from the breeding hapa, their weight is taken into consideration in view of faecal droppings. If the spawning is normal, one lakh good eggs can be available from a female of about 1 kg weight.

Next action is that the weeds with the attached eggs are brought from the breeding hapa to hatching hapa and uniformly spread there. Good eggs hatch out after about 40 hours at temperature ranging between 28 to 30°C. The period of incubation is long during the winter and short during other months.

After 4 to 5 days since spawning time, the spawn is collected from the hatching hapa in the form of tiny larvae in size of 4.0 to 5.6 mm. A better method is that the spawn from the hatching hapa is sieved through a coarse mosquito netting cloth to separate wastes or debris. The quantitative estimation of spawn is made through the use of small measuring cup of known capacity. In most cases, the number of spawn per cup is generally 400 spawn/ml. The spawn is then transferred to prepared nursery ponds or pits.

Stripping Insemination in Exotic Carps

The response of injecting pituitary extract to silver carp or grass carp is generally late; they spawn 8 to 10 hours after the second injection. A manual method of stripping has been shown in the figure given here for silver carp and grass carp (Fig. 5.7). Stripping action is required also in the breeding of common carps, as mentioned before.

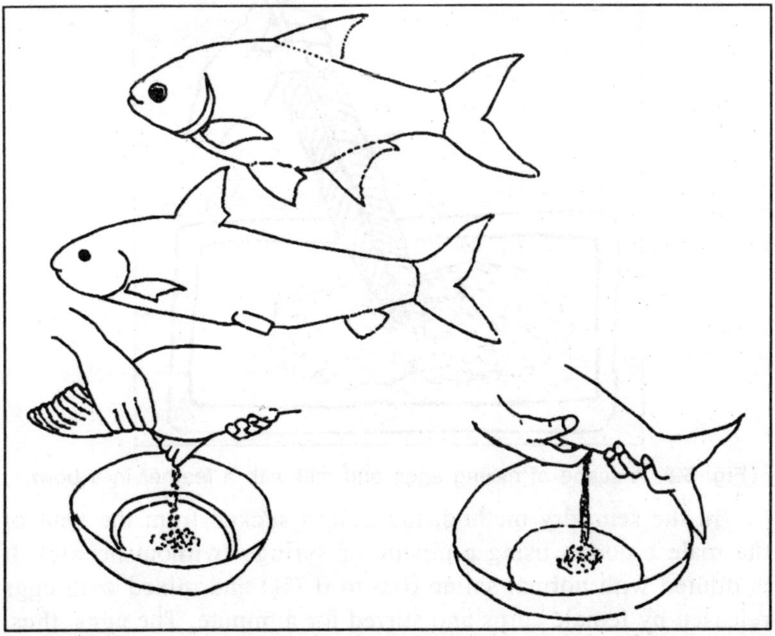

Fig. 5.7. A view of stripping action in Chinese carps (silver carp and grass carps).

There are three small well-known methods of stripping : dry method, semi-dry method and wet method. Out of these, the dry method is easy to adapt and most commonly practised. When it is marked that the female is in oozing condition, the belly is pressed gently and the eggs are stripped out into an enamel tray (alternately into a plastic basin). Male breeder with oozing condition is brought to experimental site and gentle pressure is exerted on abdomen and, thus, stripped to get release of milt in the same enamel tray or plastic basin. With the help of a clean bird's hair (alternately a plastic spoon) the milt and the eggs are thoroughly mixed (Fig. 5.8). A little saline water (0.6 to 0.7%) is added to the mixture and stirred for a minute. The mixed lot of eggs, fertilized eggs and milt (unused) is then washed several times. The fertilized eggs are given sometime treatment to get hardened in water. After doing quantitative estimation the eggs are transferred to the hatchery for hatching.

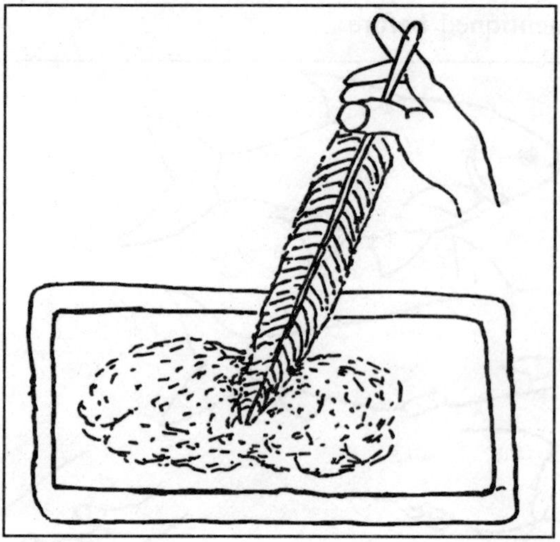

Fig. 5.8. A scene of mixing eggs and milt with a feather in a bowl.

In the semi-dry method, the milt is sucked from the vent of the male brooder, using a pipette or syringe (without needle). It is diluted with normal saline (0.6 to 0.7%) and mixed with eggs released by female carps and stirred for a minute. The eggs, thus, get fertilized.

In the wet method of stripping, the released eggs and milt are pressed out in a plastic basin having a little water. Within no time viable eggs get fertilized (Fig. 5.8).

Advantages of Induced Breeding

The process of hypophysation in carps is advantageous to farmers in many ways. It provides, in addition to augmenting seed resources, pure quality of fish seed in the fish farm site itself.

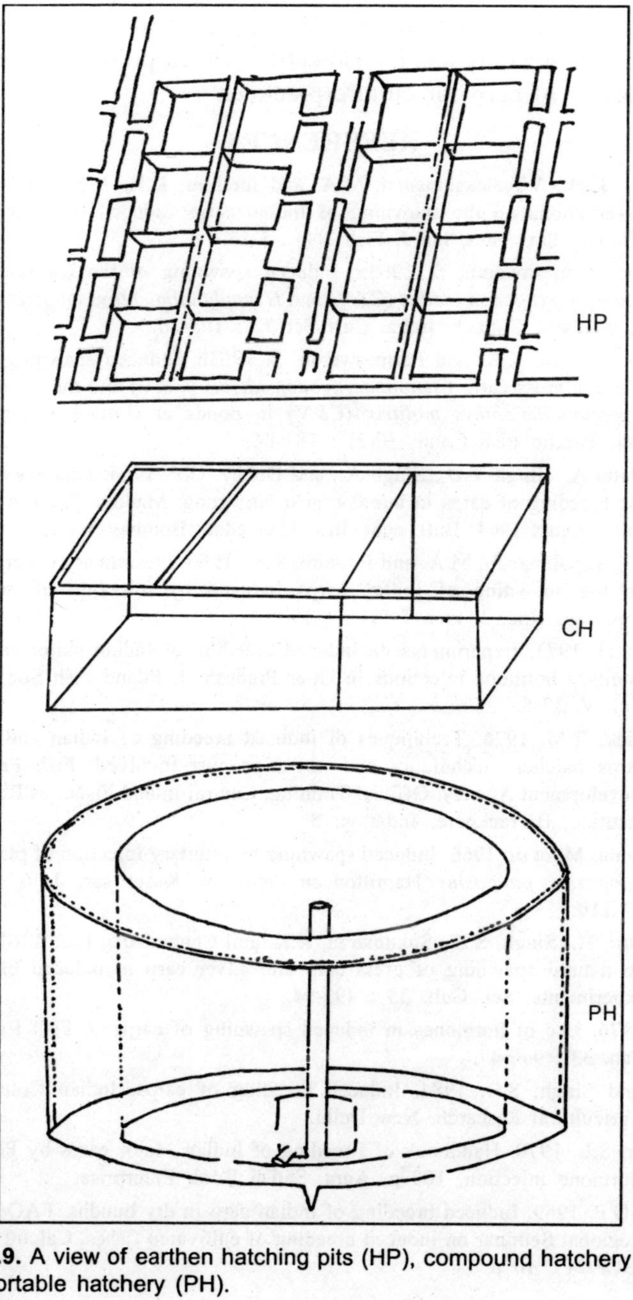

Fig. 5.9. A view of earthen hatching pits (HP), compound hatchery (CH), portable hatchery (PH).

Induced breeding technique in carp can facilitate the purpose of selective breeding and hybridization, which produce better and new variety of carp for culture practices.

REFERENCES

Alikunhi, K.H., Vijyalakshmanan, M.A. and Ibrahim, K.H. 1960. Preliminary observations on the spawning of Indian carps induced by injection of pituitary hormones. Ind J. Fish 7(1) : 1-19.

—— and Parmeshwaran, S. 1963a. Induced spawning of the Chinese carps *Ctenopharyngodon idellus* (C&V) and *Hypophthalmichthys molitrix* (C&V) in ponds at Cuttack, India. Curr Sci 32 : 103-26.

——, Sukumaran, K.K. and Parmeswaran, S. 1963b. Induced spawning of the Chinese grass carp, *Ctenopharyngodon idellus* (C&V) and the silver carp, *Hypophthalmichthys molitrix* (C&V) in ponds at Cuttack, India Proc. Indo-Pacific Fish Coun. 10(2) : 181-84.

——, Datta A., Singh V.D., Singh A., and Dubey, G.P. 1964. Observations on the breeding of carps in bundhs, near Nowgong, Madhya Pradesh during July-August 1964. Bull. cent. Inst. Fish educ. Bombay 1 : 22.

——, Vijyalakshmanan, M.A. and Ibrahim, K.H. 1960. Preliminary observations on the spawning of Indian carps induced by injection of pituitary hormones. Ind. J. Fish 7 (1) : 1-19.

Anand, J.N. 1973. Experiments on induced breeding of Indian major carps by pituitary hormone injections in Uttar Pradesh. J. Inland Fish Soc. India, Vol. V. 37-5.

Bhowmick, R.M. 1976. Techniques of induced breeding of Indian and exotic carps hatchery techniques and operation cost involved. Fish Farmer's Development Agency, Official Training, Central Inland Fisheries Research Institute, Barrackpore, India, p. 5.

Boonbrahm, M. *et al.* 1966. Induced spawning by pituitary Injection of pla-sawai *Pangasius pangasius* Hamilton in captivity. Kaset sart J. 6 (3/4) : 97-110.

Chaudhuri, H., Singh, S.B., Sukumaran, K.K. and Chakrabarti, P.C. 1967. Note on natural spawning of grass carp and silver carp in induced breeding experiments. Sci. Cult. 33 : 493-94.

——, 1976. Use of hormones in induced spawning of carps. J. Fish Res. Bd. Can. 33 : 940-47.

——, and Singh, S.B. 1984. Induced breeding of carps. Indian Council of Agricultural Research, New Delhi.

Chondar, S.L. 1970. Handbook of Breeding of Indian Major carps by Pituitary Hormone injection, 100 p. Agra. Satish Book Enterprise.

Dubey, G.P. 1969. Induced breeding of Indian carp in dry bundhs. FAO/UNDP Regional Seminar on induced breeding of cultivated fishes, Calcutta. FRI/ IBCF/21 : 10 p.

Ganapati, S.N. and Chacko, P.I. 1954. Some observations on the spawning of Indian carps in the bundhs of Bengal Indian geogr. J. 27 (3&4) : 1-17.

Kamal, M.Y., Sukumaran, K.K., Jaitley, R.N., Singh, D.N. 1979. Observation on induced spawning of grass carp, *Ctenopharyngodon idella* (Val.) and silver carp. *Hypophthalmichthys molitrix* (C&V) at Ranchi. J. Inland Fish Soc. India 11 (2) : 125-27.

Khanna, D.V. 1957. Observation on the spawning of the major carps at a fish farm in the Punjab. Ind. J. Fish 5 (2) : 282-90.

Konradt, A.G. 1968. Method of breeding the grass carp, *Ctenopharyngodon idella* and the silver carp, *Hypophthalmichthys molitrix* (C&V) FAO Fish Rep. (44) 4 : 195-204.

Lakshmanan, M.A.V., Sukumaran, K.K., Murty, D.S., Chakraborty, D.D. and Philipose, M.T. 1971. Preliminary Observations on intensive fish farming in fresh water ponds by the composite culture of Indian and exotic species. J. Inland Fish Soc. India 2 : 1-21.

Singh, S.B., Reddy, P.V.G.K., Dey, R.K., Mishra, B.K., and Muduli, H.K. 1979. Observations on prolonged induced breeding of silver carp. In Symposium on Inland Aquaculture (Abstract) February12-14, 1979 : 9 CIFRI, Barrackpore.

Sinha, V.R.P., Jhingran, V.G. and Ganapati, S.V. 1974. A review on spawning of the Indian major carps. Arch Hydrobrol. 73 (4) : 518-36.

Sinha, V.R.P., Sharma, B.K. and Chaudhuri, N.K. 1979. Breeding of silver carp, *Hypophthalmichthys molitrix* (C&V) and grass carp, *Ctenopharyngodon idella* (Val.) in a bundh type tank in West Bengal. Curr. Sci. 44 (7) : 230-31.

—— and Ramachandran V. 1985. Fresh water Fish Culture. Indian Council of Agricultural Research, New Delhi.

6

Fish Seeds

For fresh water fish culture practices, there is need of application of proper technique and perfection in the management of pond in respect to the use of organic and inorganic fertilizers, availability of planktons, utilization of these and some other aquatic vegetation and animal organisms by fishes, eradication of unwanted weeds, pests and predators, etc. However, stocking of pond with the fast growing fish seeds of high grade quality is the most precious requirement.

Unfortunately the cultured species of major carps, commonly known for their very fast growth and other qualities, do not ordinarily breed in ponds. In this situation, the seeds have either to be collected from the flooded rivers where these carps spawn once a year during the monsoon season or produced through artificial means, such as bundh breeding or hypophysation. Whatsoever the numerical figure or volume, the dependence on riverine fish collection of seed is now very much limited. The main reasons for this non-preference are mixed quality with doubtful trait, low availability and production uncertainty. This situation has led to dependence on other methods such as (i) bundh breeding, and (ii) induced spawning by hypophysation. Seed production from bundhs, specially dry bundhs, is reliable in quality and, therefore, preferred. The yield is 100% of pure quality. The process is simple and in a single operation, mass production of spawn is available. The production of seed from bundhs has surpassed even the riverine spawn during last decade and now 70% of the major carp spawn for sale at Howrah Fish Market are those from bundhs. From records it is evident that the fish seed production during 1995-96 has been 15,007 million fry with a target of 16,500 million fry during 1996-97.

Natural Resources of Fish Seeds

Riverine collection of spawn of major carp still remains as a source of spawn and is used on commercial scale for culture of carps in Bihar, West Bengal and Uttar Pradesh. However, Bihar alone fulfils 51% requirement of the country's total production of fish spawn and fry.

Riverine Collection

The seed of major carps, *i.e.*, of *Catla catla, Labeo rohita* and *cirrhinus mrigala* and of some other carps as well are collected during the monsoon (August to October). During this period, the mature fish ascend the flooded water of the rivers and indulge in breeding action on finding suitable ecological and meteorological conditions. In the Ganges river *i.e.,* mostly in Bihar part of this big river, J.S. Datta Munshi *et al.* (1979) observed that the fish seed collection grounds were located at Brauni, Hathidah, Monghyr (Lankarghat), Sultanganj, Akbarnagar, Sonbarsa, Bhagalpur, Colgong, Sahebganj, Rajmahal, Manikchak and Farakka (West Bengal). It was marked by them that the collection grounds were usually confined between erosion zones and shallow zones with a moderate current of water (Fig. 6.1).

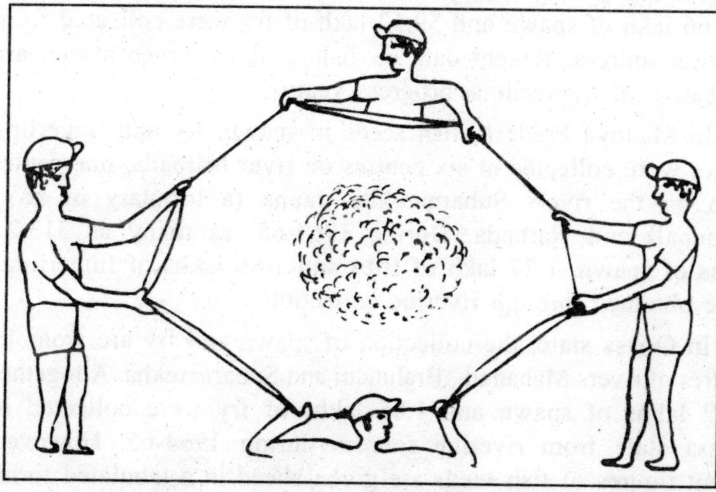

Fig. 6.1A. Fish seeds placed in a cloth, shifted from hatching hapa.

In respect to spawn resources, Bihar is the most important state covering the greater percentage of the country's total collection of fish seeds in the form of fish spawn from reverine sources. Besides Ganges, river Sone, river Kosi, river Gandak and their tributaries are the major sources. As per estimation, the Ganges spawn generally comprise 75% major carps, of which mrigal are 50% and catla and rohu are 25% each. The Sone spawn generally includes 50% major carps, of which mrigal are about 60%, Rohu 25% and Catla 15%. The Kosi collection of spawn generally includes 50% major carps and the rest others.

During 1964-65, the collection from rivers comprised 1000 lakh spawns. Assam has poor collection of seeds from natural sources. River Brahmaputra and her tributaries constitute the natural fish seed resources. The quality of seed collected from rivers is not good for the reason that the survival potentiality is not high. In Uttar Pradesh, the spawn of major carp in good form are collected from several sites of riverine sources. It has been reported that the spawns are collected at about 150 centres on the rivers Betwa, Chandmovinala, Ghagra, Gomti, Ganges, Yamuna, Hinden, Isun, Krishna, Ramganga, Rapti, Sahnadi, Sai, Sarda, Sajoo and Sone. The fry are collected at eight centres on the rivers Ganga, Ghagra, Sarju and Sone. During 1964-65 altogether 821.66 lakh of spawn and 59.12 lakh of fry were collected from riverine sources. Recent data on fish seeds as given ahead, are indicative of tremendous progress yearwise.

In Madhya Pradesh, fish seeds in spawn, fry and fingerling stages were collected at six centres on river Narbada, one centre each on the rivers Subarnarekha, Jauna (a tributary of river Chambal) and Narbada. During 1964-65, as many as 313.22 lakhs of spawn, 1.77 lakh of fries and 1.86 lakhs of fingerlings were obtained through riverine collection.

In Orissa state, the collection of spawn and fry are from 15 centres of rivers Mahanadi, Brahmani and Subarnarekha. Altogether 1342 lakhs of spawn and 1.05 lakhs of fry were collected in Orissa state from riverine sources during 1964-65. However, recent figures of fish seeds are given ahead in a tabulated form.

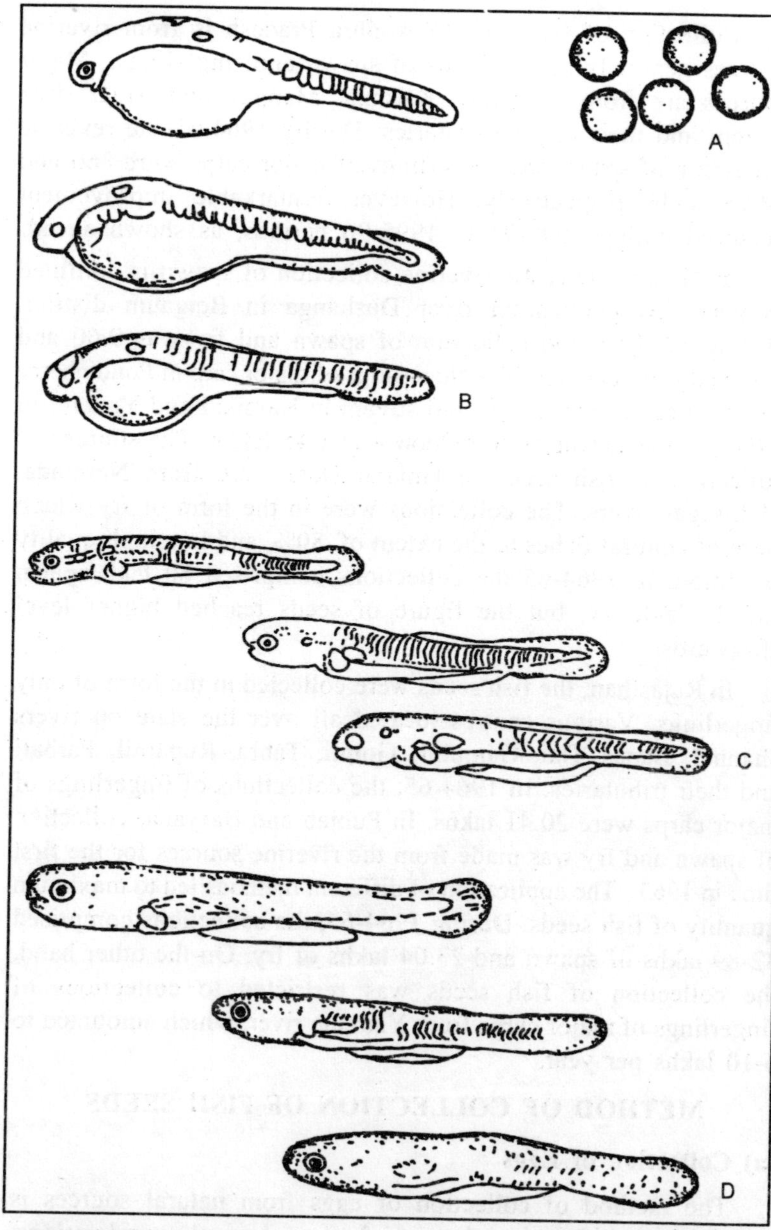

Fig. 6.1B. Early forms of fish seeds: Eggs before hatching (A), hatchlings of 3 major carps (B), Fries of 48 hours (C), the same of 72 hours (D).

Collection of fish seed in Andhra Pradesh is from riverine sources generally in the forms of spawn and fingerlings. Major sources are from Godavari, Krishna, Manjera and Muai river system and their major tributaries. During 1964-65 the reverine collection of spawn and fingerlings of major carps were 360 and 24.75 lakhs respectively. However, remarkable improvement occurred during 1990-91 to 1995-96 periods, as shown ahead.

In Mysore state, the riverine collection of spawn is confined to very few centres on river Dushanga in Belgaum district. During 1964-65, the collection of spawn and fry was 9.60 and 3.91 lakhs respectively. The fingerlings of major carp in Pondicherry were collected from rivers and stream in Karaikal and Yanam. In 1964-65, the extent of collection was 1.45 lakhs. The sources of collection of fish seeds in Gujarat state were from Narmada, Mahisagar rivers. The collections were in the form of fry which were of cultural fishes to the extent of 80%, and in a high quality condition. In 1964-65 the collections comprised 80 lakh spawn and 10 lakh fry, but the figure of seeds reached higher level afterwards.

In Rajasthan, the fish seeds were collected in the form of only fingerlings. Various centres located all over the state on rivers channel, Banas, Dai, Gambhir, Gomti, Tabra, Ruparail, Parbati and their tributaries. In 1964-65, the collections of fingerlings of major carps were 20.41 lakhs. In Punjab and Haryana, collection of spawn and fry was made from the riverine sources for the first time in 1963. The application of different methods led to maximum quantity of fish seeds. During 1964-65, the collections comprised 32-89 lakhs of spawn and 23.04 lakhs of fry. On the other hand, the collection of fish seeds was restricted to collections of fingerlings of major carps from Yamuna river, which amounted to 6-10 lakhs per year.

METHOD OF COLLECTION OF FISH SEEDS

(a) Collection of Eggs

The method of collection of eggs from natural sources is based on scientific principles, mainly in ecological considerations. After locating actual breeding grounds, the eggs are scooped out from shallower grounds by means of rectangular piece of mosquito netting of varying sizes, having fine meshwork of nylon.

From fast flowing waters, the eggs are collected by Benchi Jal, which is a kind of conical bag net (Fig. 6.2). Formerly special technique was applied for the collection of carp eggs at Naizora in river Dhikhow in Sibsagar (Assam), which is a fast flowing river. As there was much difficulty in the collection work and the collected item was not sufficient due to high current in the river, this practice has been almost abandoned. The need has been fulfilled by taking up modern induced breeding technique in which cost involvement is much less and quality of seeds and their numbers obtained are of high order.

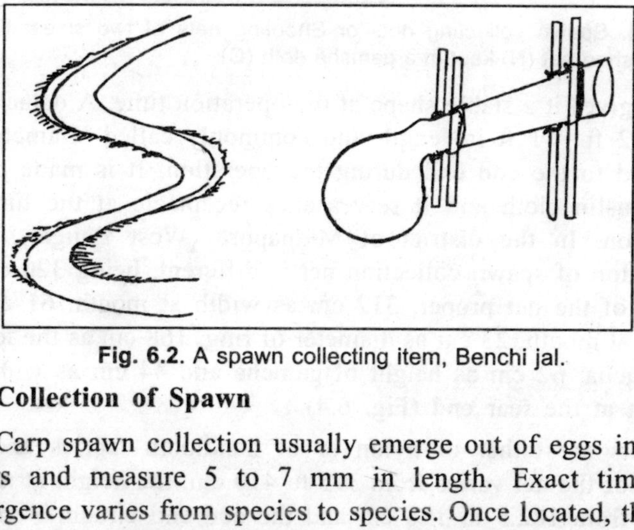

Fig. 6.2. A spawn collecting item, Benchi jal.

(b) Collection of Spawn

Carp spawn collection usually emerge out of eggs in 18-24 hours and measure 5 to 7 mm in length. Exact timing of emergence varies from species to species. Once located, the carp spawns are collected with the help of a specially devised net, called Benchi jal in Assam and Bengal. It is a sort of shooting net (Fig. 6.3). The design, size and construction material of the Benchi jal differs in different areas and state. The larger ones were conical in shape and open at both ends, measuring about 24 feet in length. They possessed a wide mouth with a diameter of about 18 ft. The latter may be provided with two lateral wings to widen the receptive area of the mouth. The cod end measures 9″–12″ in diameter and is provided with a cane or bamboo ring

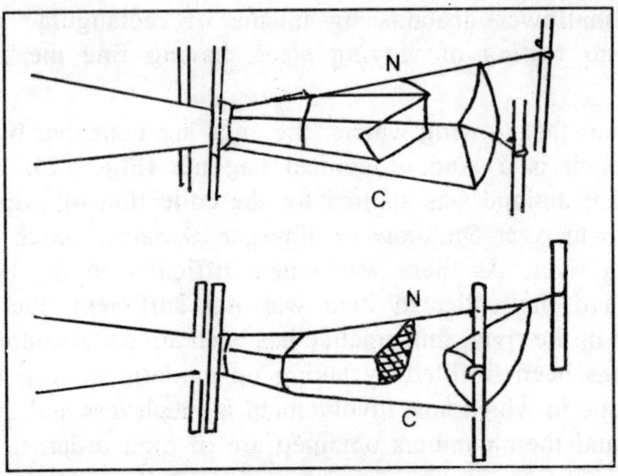

Fig. 6.3. Spawn collecting nets or Shooting nets of two sizes: Round meshed net (N) kept in a gamcha doth (C).

which gives it a stable shape at the operation time. A detachable piece, 2 ft × 1 ft in length and commonly called 'Gamcha' is attached to the cod end during the operation. It is made up of fine muslin cloth and it serves as a receptacle at the time of operation. In the district of Midnapore (West Bengal), the dimension of spawn collection net is different, being 320 cm in length of the net proper, 312 cm as width at mouth, 61 cm as height at mouth, 23 cm as diameter of ring, 168 cm as the length of gamcha, 62 cm as height of gamcha and 44 cm as width of the net at the rear end (Fig. 6.4).

Nets are either of nylon or of handloom fabrics and the length of the net varies from 395 to 410 cm, the height at mouth varies from 182.5 to 196 cm and the cod end circumference is between 64.0 to 70.0 cm. The number of meshes in one square centimeter is between 100 to 144.

Method of net operation. The fishermen generally operate the spawn collecting net in shallow margins of the flooded river, keeping the mouth of the net facing the current. The net is fixed to bamboo poles in shallow water with low current, the mouth facing the current but its upper lip is kept a little above the water level. The cod end is kept drifting in the diameter of the current in which position it is fixed to poles just below the water surface.

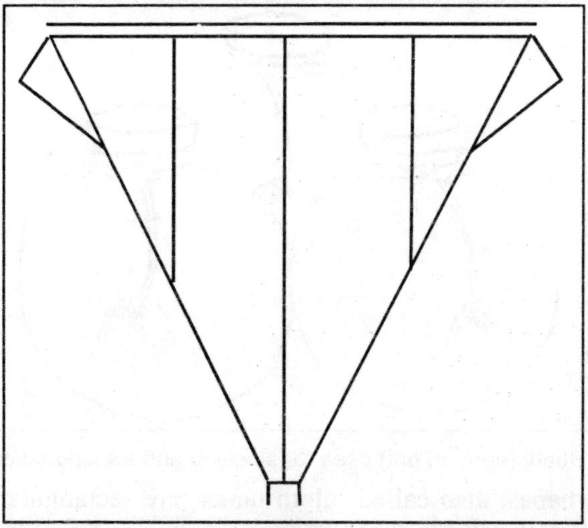

Fig. 6.4. A view of Midnapore type of spawn collecting net (not as per scale).

To this end, the gamcha is properly tied and its hind end is tied to another pole. In this position, the drifting spawn along with current pass into the net and get accumulated in the bag, *i.e.*, gamcha from where the spawns are periodically removed and stored in small pits or hapas after collecting them first in 'hundies' and subsequently transferring them. The latter, also called 'cloth tanks' are 'rectangular' in shape and made of fine muslin cloth (Fig. 5.5). The spawn collecting nets are arranged in series as per depth and force of the water current. In shallow depths of 125 to 165 cm, such nets upto 7 numbers are used.

The 'hundies' are ordinary earthenware pot with ample space for temporary storing fish-spawn and keeping the water cool (Fig. 6.5). A common practice is that a red coloured earthen substance, hill mitti, is added to the content of hundies. Its speciality is that it maintains pH, renders water turbid after getting dissolved. The loss of transparency in water disallows penetration of sunlight. Subsequently, the spawn in the hundies get settled.

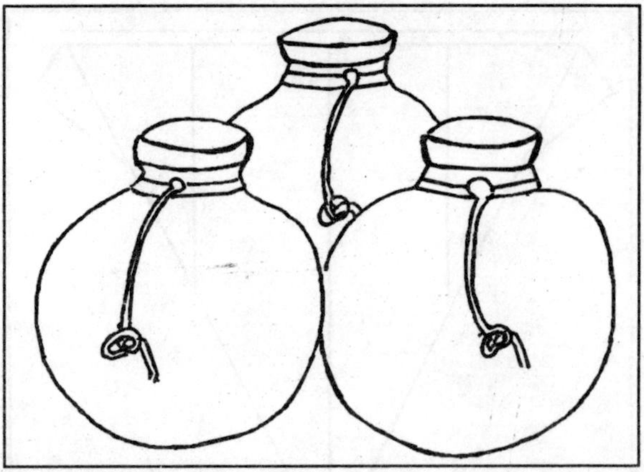

Fig. 6.5. Hundi (earthen pot) used for stocking and transportation of seeds.

The hapas, also called 'cloth tanks' are rectangular in shape and are fixed with the help of four bamboo poles in such a way that its upper edges always remain a few inches above the water (Fig. 5.5). Care is also taken to keep the hapa free, not resting on the bottom mud; also its bottom part does not float in water due to the presence of air bubbles.

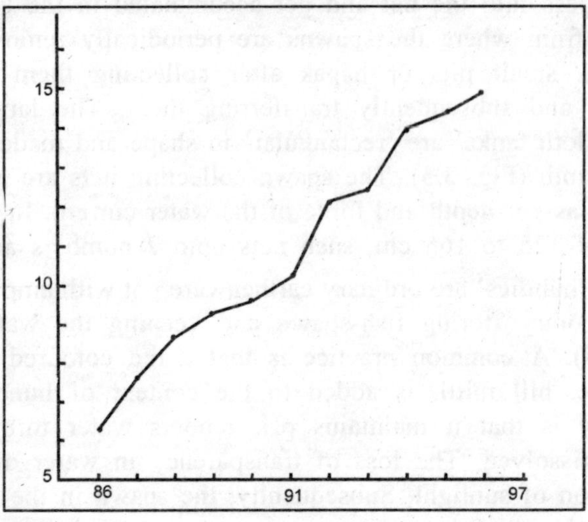

Fig. 6.6. Graphic presentation of seeds (fries) production for 10 years (1986 to 1997). Vertical line stands for number of seeds in millions and horizontal line for yearwise events.

(c) Collection of Fry and Fingerlings

Though collection of spawn is most favoured practice in the Gangetic river and its tributaries in some places (Fig. 6.7), a prevalent practice is the collection of fry and fingerlings from the irrigation canals and paddy fields (David, 1959; Chaudhuri and Singh, 1984).

In some states, such as Punjab and Madhya Pradesh, the fishermen collect fingerling from the irrigation canals or pick-up weirs and in Andhra Pradesh and Tamil Nadu, the fries and fingerling are collected also from the paddy fields. They generally use small nets, which are fine meshed drag nets, for their collection. But for the collection of fingerlings, cast nets, traps or fine muslin cloth are also used in the positions where the young ones try to cross the irrigation barriers. In the rivers of Godavri, Krishna and Kavery, basket traps are commonly used for collecting fry and fingerlings during monsoon.

Fig. 6.7. A scene of collection of fish seeds from a river.

Measurement of Spawn

The measurement of spawn quantity is possible through indirect method. The live spawn from hapas is removed by muslin cloth, allowing water to drain out. The latter is brought to special siege cups having wire-meshed (1/24″ mesh) bottoms. Measuring cups of various capacities are used, such as 5 ml, 10 ml, 20 ml, 30 ml, 50 ml, 100 ml and 200 ml. The average size of a spawn is 5.5-6.5 mm. With this measurement in consideration, the average number of spawn in a cup of 5 ml is approximately 750/ml. However, in a larger cup it is about 500/ml.

Quality of Riverine Collection of Spawns

From river sources, live-spawn is collected in many places of West Bengal and Bihar. As spawn is a highly perishable commodity, it needs disposal (*i.e.*, marketing) within two days. The spawn from known sources and also the timing of collection are important consideration's for ascertaining marketing value. If the availability of fish spawn, *i.e.*, seeds produced by induced breeding or bundh breeding is at the same time as the riverine spawn, the market prices and, therefore, margin of profitability are significantly influenced. Moreover, the natural variations in flood pattern and monsoon arrival very much affect the production of spawn. As such a risk is involved in respect to quality and quantity of spawn at a centre of collection. The margin of profit from marketing of riverine spawn collection is highly speculative and subject to many limitations. In spite of these odds, there is even now a flourishing trade of fish seed at Howrah Centre. The Gangetic system in Bihar and upper stretches of the Hooghly in West Bengal still remain the main collection centres of riverine spawn. Following considerations are most vital to boost up the fish production :

(i) Improvement in gear technology.

(ii) Better gear operation.

(iii) Easy and quick modes of transportation of seeds from one place to other, *e.g.*, marketing centre.

(iv) Provision of more infrastructural facilities to enable quick supply of spawn, *i.e.*, fish seeds to marketing centres.

Information on Production of fish seeds (fry) is available which is given in table 6.1 for top ten states. It is evident that order of fish seed (fry) production in the top ten states, from highest to lowest, is respectively as West Bengal, Assam, Andhra Pradesh, Madhya Pradesh, Uttar Pradesh, Tamil Nadu, Bihar, Maharashtra, Haryana and Orissa. All India figure for seed production in 1990-1996 is respectively as 10332.23 in (1990-91), 12202.91, 12498.91, 14218.93, 14544.19 and 15007.22. A yearwise status for six years is given in Table 6.1 (for 10 top fish seed producing states).

Table 6.1

Fish seed production figures of top ten states during 1990-96 (Quantity in million fry)

States	1990-91	1991-92	1992-93	1993-94	1994-95	1995-96	% of Total in 1995
Andhra Pradesh	452.20	490.00	600.00	600.00	700.00	709.00	4.72
Assam	193.64	1216.00	1626.00	2170.81	2386.95	2547.54	16.97
Bihar	347.63	355.90	328.23	360.23	321.16	332.21	2.21
Haryana	300.00	339.00	334.52	459.93	523.40	564.20	3.75
Maharashtra	75.51	256.49	299.97	322.00	270.00	293.00	1.95
Tamil Nadu	231.23	219.21	224.185	500.24	536.59	467.43	3.11
Uttar Pradesh	210.10	296.68	408.18	458.22	479.15	546.62	3.64
West Bengal	7551.00	7884.00	7539.95	8105.32	8126.14	1880.00	54.50
Orissa	202.70	214.89	198.90	185.40	188.67	186.69	1.24
All India Figure	10332.23	1202.91	12498.93	14218.93	14544.19	15007.22	

Source : Handbook of Fisheries Statistics, 1996, Govt. of India.

The All India figures of fry production for 1985-86 to 1996-97, have been given in Table 6.2 and the trend of progress has been made explicit through graphic presentation (Fig. 6.6) as given before.

Table 6.2

Year	Figure of fry
1985-86	6322 Million fry
1986-87	7601 ”
1987-88	8608 ”
1988-89	9325 ”
1989-90	9691 ”
1990-91	10332 ”
1991-92	12203 ”
1992-93	12500 ”
1993-94	14239 ”
1994-95	14544 ”
1995-96	15007 ”
1996-97 (Target)	16500 ”

Source : Handbook of Fisheries Statistics, 1996, Government of India.

7

Transportation of Fish Seeds

For a successful fish culture practice, search for the solution of the problem of transport of live fish spawn cannot be overlooked or ignored. There is likelihood of mortality of live fish spawn during and after transport due to the following reasons (in order of effectiveness) :

(i) Oxygen depletion, followed by starvation of blood and tissues.

(ii) Containment of some toxin in the transport-water media.

(iii) Hyperactivity of spawn leading to strain and exhaustion.

(iv) Diseases infested during transport.

(v) Infliction of injuries due to jolts or as a result to overcrowding and starvation.

(vi) Doubtful quality of water media.

Role of Oxygen. Oxygen concentration in the water media influence the viability capacity to a great extent. The individuals of spawn stock consume oxygen and remain in active form. The mere presence of sufficient level of dissolved oxygen in water does not necessarily indicate that the blood and tissues are sufficiently charged with oxygen. There may be hindrance in this charge, *i.e.*, ability of haemoglobin to combine with oxygen, due to existence of low pH, high carbon dioxide and ammonia level in water. Rise of temperature accelerates the metabolic rate and rate of oxygen consumption also (Khan, 1946; Singh, 1977).

Effects of containment of toxin. Brakway (1950) has explored that fish are sensitive to both ammonia and ammonia compounds. Incidentally large quantities of ammonia are excreted by fish. Also bacterial decomposition of urea and other nitrogenous wastes, slime and dead fry contribute to production of ammonia, sometimes to an alarming level.

Bacteria, plankton and other organic matters may be present in the packing of spawn. These consume oxygen and produce toxic end-products. Keeping these possibilities in view, the spawn should be packed in a clear, well aerated, bacteria-free or toxin-free water.

Hyperactivity effects. When spawn is packed in a container and transported in oxygen-rich water media, it may also happen that after handling and transport jerks received, the spawn may react out of excitement and die during transport. This is the main reason that use of some suitable sedative has been recommended to quieten the spawn.

Infestation of disease. The water media in which transportation of spawn is done, may be contaminated with some infective germ, or some individual of the spawn may be having internal infection from the site of their collection. In the crowded condition of spawns in a pack, the internal infection of any individual may get easily transmitted to other individuals and, after sometime, the disease may create havoc by causing mass mortality.

Infliction of injuries. In a crowded condition inside a pack and in a stage of transportation, some of the over excited individuals may get inflicted with external injuries and death may occur consequently. The injurious condition may facilitate bacterial or fungal infections.

Doubtful quality of water media. The use of a clear, tested to be contaminant-free water, is a must for a successful transportation of the fish spawn. Any lapse in prior check up of water quality may defeat the purpose of transportation.

Threat to viability is greater in younger stages of fish seeds than bigger ones. It has been reported that the experienced chinese fresh water fish farmers prefer to transport fish seeds to long distance only in a size of the fry larger than 40 mm.

Method of packing and transport. There are two well-known methods of packing and transport in practice.

(i) in open carrier with or without provision of artificial aeration/oxygenation/water circulation (Fig. 7.1).

(ii) in sealed air-tight carriers with oxygen.

Fig. 7.1A. A view of hundies (earthen pots) containing fish seeds ready for transportation by fishermen on sling.

Open carrier system. Traditional method of fish seed transportation was in earthen hundles, half filled with the river water and about 10,000 to 15,000 fish spawn were kept for transportation (this amounted to about 30 ml figure). Afterwards, this was replaced by aluminium containers. However the hundies are in use in some parts of eastern India even now. They are manually carried on slings or on bicycle or by boat to marketing place (Fig. 7.1). The usual practice is that slight jerk and splashing of water is given at times to the carrier for aeration of the surface water and enriching supply of dissolved oxygen to the spawn (in carrier). In this system, the transportation is possible to only short distances, as manual labour is strainous. Moreover, in this crude practice, there is generally 30-40% mortality of spawn, if the transportation is for a fairly long distance. Addition of fine silt or 'red soil' allows coagulation of the suspended organic pollutant. If necessary, water is partially removed or replaced by fresh water. Galvanized tin container of

40-50 litre capacity, mounted on a wooden base, prevent heating of the container by contact with the ground. A perforated disk-like insert lid is suitably fitted which serves for aeration through splashing action with water. After sometime the used water is allowed to be replaced by fresh one to provide enrichment of the dissolved oxygen in water.

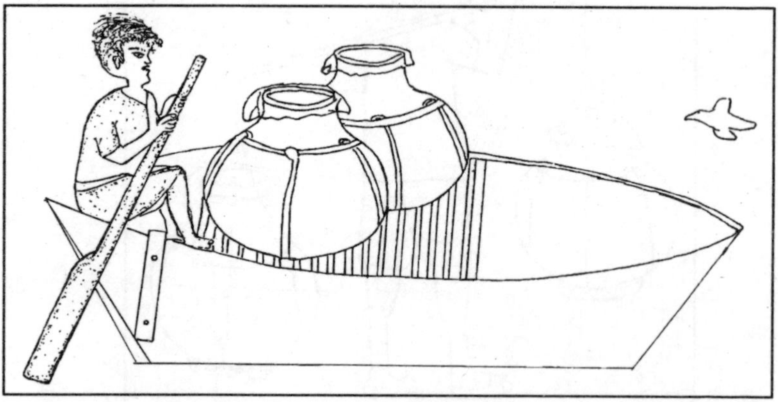

Fig. 7.1B. A view of transportation of fish seeds by boat.

In some parts of India the fingerlings are transported by open tanks mounted on a truck and provided with mechanical facilities of aeration. Open canvas containers in size of 1 m × 1 m × 1.25 m are in common use in Punjab and Madhya Pradesh for the transportation of fingerlings. The carriers of spawn in some foreign countries of South-East Asia are of different nature. In China, bamboo baskets and wooden tubs are used for transportation, whereas in Indonesia, the carrier is in the form of water tight, tar-coated, plaited bamboo basket of 15 litres capacity to accommodate 10,000 fry of average 50 mm size. It is considered as an economical device. The Chinese baskets or tubs have maximum height of 2 metre and diameter of 2 metre. In Philippines and Indonesia, there is use of boats for transportation of fish fry. The boats are designed to allow free water circulation in them. Some of these boats are so built that they have loading capacity of 10 million fry.

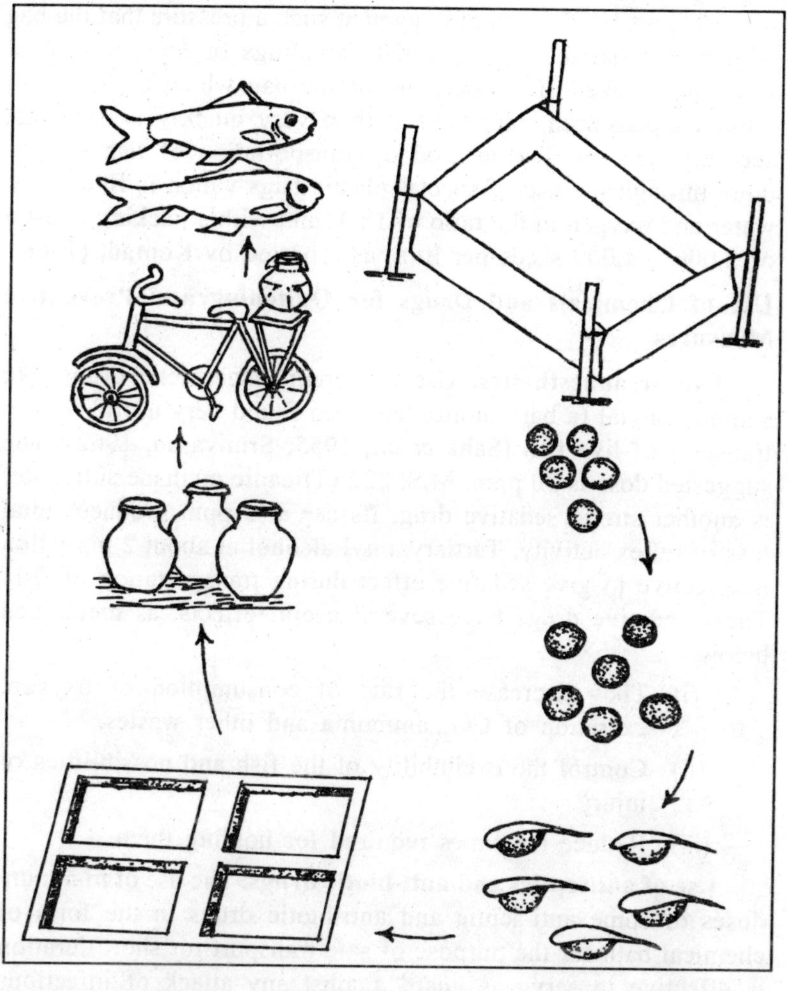

Fig. 7.2. A view of production of seeds and another method of easy transportation.

Closed system. The most significant development in the line of transportation of fish seed is the method of packing live fish seeds under oxygen for transportation. In this technique, plastic bags are used with sealed atmosphere of oxygen for transportation of fish fry. This was introduced in India in 1955 and it gained popularity in a short time in this country. The method of use is like this : In a thin plastic bag (300 gauge) with about 6-7 litres of water and 10-12 litres of oxygen at such a pressure that the bag is lightly inflated, 3,000 – 6,000 hatchlings or fry per litre of water are packed. The open end of the bag which is kept tight around the oxygen inlet tube is then very quickly twisted and securely tied up. The method of transportation of fish seed is done through the use of special plastic bags which is filled with water and oxygen in the ratio of 1 : 1, and with a packing density of 2,000 – 4,000 seeds per litre, as reported by Konradt (1968).

Use of Chemicals and Drugs for Quitening and Preventive Measures

Use of anaesthetics. Use of certain anaesthetic drugs like sodium amytal (a barbiturate) has been found very useful for the transport of live fish (Saha *et al.*, 1955; Srinivasan, 1962). The suggested dose is 50 ppm. M.S. 222 (Tricanie methane sulfonate) is another strong sedative drug. Its use at 5 ppm produces total loss of reflex activity. Tertiary amyl alcohol at about 2 ml/gallon is effective to give sedative effect during transportation of fish. These sedative drugs have several useful effects, as mentioned below :

(i) They decrease the rate of consumption of oxygen, excretion of CO_2, ammonia and other wastes.

(ii) Control the excitability of the fish and possibilities of injury.

(iii) Reduce the times required for holding them.

Use of antiseptics and anti-biotic drugs. The use of minimum doses of some anti-septic and anti-biotic drugs in the form of chemical bath for the purpose of safe transport for short duration is effective to serve as guard against any attack of infectious diseases or parasites. Meyer (1968) and Gopalkrishanan (1968) and some others have recommended the use of some chemicals in specific concentrations as given below :

Acriflavine	10 ppm
Methylene blue	2 ppm
Copper sulphate	0.5 ppm
Potassium permanganate	3 ppm
Chloromycetin	8-10 ppm

Sodium chloride	3 per cent
Formalin	15 ppm

The use of any of these chemicals (as alternatives to one another) saves the fish seed during transport. However, expert advice is that these chemicals should not be added in the transport water, but sodium chloride is an exceptional in transporting fingerlings and adults. This chemical gives another advantage : it controls the strain of osmotic pressure by making the body water isotonic to the transport water and, in this way, saves the energy of the fish seed, liable to be spent in adjusting to the surrounding environment.

Mc Farland (1960) has reported that use of an organic buffer, Trihydroxy-methyl amino methane, at a dose of 5-10 gm/gallon *i.e.*, 1.302.6 gm/litre, is effective both in fresh water and sea water to control pH and CO_2 levels and, thus help in the viability of the fish seeds during the transport to a marketing centre or nursery site.

Additional Specific Suggestions for Safe Transport

(1) The ideal time of capture and handling of fish seeds during collection is early morning, as the water media, then, is cooler and in calm condition.

(2) The fish seeds (fry or fingerlings) should be allowed acclimatisation for at least few hours in the new water media, which is destined for use in transportation.

(3) The density (number/ml volume of water) of the fish seed to be transported should be estimated before hand to allow sufficient dissolved oxygen during transit. Keeping this point in consideration, the spawn, fry or fingerling should be packed in separate containers.

(4) The container or carrier should be painted white on outer surface to protect against heat effects.

In consideration to various precautionary measures and application of techniques, the annual production figure of fish seeds in 1994-1995 had reached 14544 million figure, and the same in 1995-96 had reached 15007 million figure, whereas in 1996-97, the largest was 16500 million. The all India figures for

1985-1986 to 1996-1997 have been given in Table 6.2 in the earlier chapter and the trend of progress has been made explicit through a graphic presentation given before (Fig. 6.6).

REFERENCES

Basu, S.P. 1951. Effect of addition of soil in transport on larvae of Indian carps. Proc. Ind. Sci. Congr. 38(3) : 227.

Chaudhuri, H., Singh, S.B. and Sukumaran, K.K. 1966. Experiments on large scale production of fish seed of the Chinese grass carp, *Ctenopharyngodon idellus* (C & V) and the Silver Carp *Hypophthalmichthys molitrix* (C & V) by induced breeding in ponds in India. Proc. Ind. Acad. 63B (2) : 80-95.

Chaudhuri, H., 1984. Induced breeding of carps. Indian Council of Agriculture Research, New Delhi.

Chen, C.S. and Lin, S.Y. 1935. The fish fry industry of China. Bull. Chekiang Prov. Fish Exp. Stn. 1 (4) : 1-50.

Datta Munshi, J.S., Bilgrami, K.S., Saha, M.P. and Johar, M.A.O. 1979. Survey of Fish spawn collection grounds in the River Ganges. Environmental conservation, vol. 6(2) : 153-57.

David, A. 1959. Observations on Some Spawning grounds of the Gangetic major carp with a note on carp seed resources in India. Indian J. Fish 6(2) : 327-41.

Dubey, G.P. and Arjun Singh, 1968. Fish Seed production from dry bundh and its economics. Seminar on production of quality fish seed for fish culture, Barrackpore. Nov. 1 & 2, 1968 (Abst.).

Dwivedi, S.N. 1983. Fish seed production : National Level Technology. In "Fisheries development in India" (Ed. U.K. Srivastava and M. Sharma Reddy. Naurang Rai Concept Publishing, H-13, Bali Nagar, New Delhi.

Ganapati, S.V. and Chacko, P.I. 1954. A comparative study of the transport of fish seed in oxygen tin carriers and in ordinary tin carriers. Ind. Com. J. 6 : 364-87.

Gopalkrishnan, V. 1968. Diseases and parasites of fishes in warm water ponds of Asia and the Far East. F.A.O. Fish Rep. (44) 5 : 319-44.

Ibrahim, K.H. 1961. Some observations on the carp fry resources of the river Godavari near Rajahmundry, Andhra Pradesh. Indian J. Fish 8(2) : 403-12.

Khan, H. 1946. Transport of alive fish in oxygenated containers. Curr. Sci. 15(2) : 51-52.

Khanna, D.V. 1957. How to get rid of aquatic insects from fish nurseries. Ind. Fing 7 (3) : 12-14.

Konradt, A.G. 1968. Methods of breeding the grass carp, *Ctenopharyngodon idella* and the silver carp, *Hypophthalmichthys molitrix*. FAO Fish Rep. 44(4) : 195-204.

Malhotra, J.C., Shetty, H.P.C. and Ghosh, K.K. 1968. Fish Spawn resources of River Yamuna. Seminar on Production of quality Fish-seed for Fish Culture. Sponsored by ICAR at CIFRI, Barrackpore, India. 1-2 November, pp. 124-59.

Mammen, T.A. 1962. Live Fish transport using modified splashless carrier with compressed in aeration. In Training Course on live fish transport. November 19-21, Hyderabad. Fisheries Extension Unit. Government of India : 5 p.

McFarland, W.N. 1960. The use of anaesthetics for handling and transport of fish. Calif Fish Game 66 (4) : 408-31.

Rahman, Q.M. 1946. A note on the fish-fry trade in Bengal Bull Directorate of Fisheries, Bengal, Calcutta : 37 pp.

Ramachandran, V. 1969. Transport of spawn, fry, fingerlings and breeders. FAO/UNDP Regional Seminar on Induced Breeding of Cultivated Fishes. Calcutta : FRI/IBCF/16 : 28 p.

Ranade, M.R. and Kewalramain, H.E. 1956. Experimental air shipment of carp fry in plastic bags. Ind. J. Fish 3 (2) : 290-93.

Saha, K.C., Sen, D.P. and Banerjee, S.M. 1955. Use of Sodium amytal in the transport of fry of Indian major Carps. Ann. Biochem. 15(2) : 133.

——, Sen, D.P. and Mazumdar, P. 1956a. Studies on the mortality in spawn and fry of Indian major carps during transport. Pt. 2. Effect of oxygen pressure, free surface area, water volume and number of fry in the medium of transport. Ind. J. Fish 3 (1) : 127-34.

——, —— and —— 1956b. Studies on the mortality in spawn and fry of Indian major carps during transport. Pt 3. Effect of inimical substances from decomposition of metabolic products in the medium on spawn life and their control. Indian J. Fish 3 (1) : 135-40.

Singh, B.N. 1977. Oxygen consumption and the amount of oxygen required for transport of rohu and mrigal fingerlings. J Inland Fish Soc. Ind. 9 : 98-104.

Srinivasan, A. 1962. Use of tranquilizing drugs and narcotics in reducing oxygen consumption by fish. Indian J. Fish 9 A (2) : 738-45.

8

Fish Farm Construction

Fish culture is mainly practised in ponds, tanks and cages. For getting accomplishment of a high aimed target of fish production, adequate knowledge of principles of freshwater fish culture and of their application is essential. There should be sufficient information of the needs of fish themselves. Prime needs of fishes are availability of (i) sufficient oxygen, and (ii) food and space so that they can have normal life activities, *i.e.*, growth and reproduction.

Fish culture practices are undertaken in slightly different ways in different kinds of water bodies. Rural ponds are commonly used by village people for farming. These ponds differ widely in shape, size, topography, soil and water qualities, water supply provisions, natural food availability and so on. The village farmers generally have enough domestic consumption of fish but the harvest from these pond had only marginal chance of profit from marketing meagre surplus fish. However, the opportunity of sufficient yield and earning sizable profit gets enhanced on following principles of fish culture, as commonly seen in West Bengal. Whatsoever, the experiences gained from these village ponds have contributed enough to develop precised techniques of fish culture. The results of experimental works by fishery scientists have brought pisci-culture in line of agricultural industry in consideration to investment of money on fish basic requirements and getting cash return through the sale of harvested fish.

One of the prime necessity of pisci-culture practices with commercial aim is the construction of suitable fish farm with all infrastructural details and provisions as per formulated technique. A full attention to meet all requirements can brighten the scope of reaching target of production.

Topography

It is the placement of pond, *i.e.,* selection of site of the pond in respect to its geographical surroundings. Foremost requirement in this connection is the design work with the consultation of fishery expert. Following this, survey is to be made for a suitable site to suffice the need as per design. Ideal suggestion is that the pond should be located in a saucer shaped area, should have a slopy bottom and the depth of the pond should be regular. Obviously the hilly area should be avoided for the construction of a new pond. The height of the dykes should be determined on the basis of land elevation and the fluctuation of water level. If a big farm is proposed with several small ponds of different sizes and depths, the total area with the main pond and subsidiaries (nursery, rearing, stocking, breeding units) should preferably be as follows :

1. Nursery ponds. It should be small (50 × 50 × 6) and shallow with 1.0 to 1.5 metre depth of water.

2. Rearing ponds. It should be larger (90 × 30 × 6) and water depth should be 1.5 to 2.0 metre.

3. Stocky ponds. It should be further larger (300 × 90 × 6) or 90 × 30 × 6) and deeper to retain 2.0 to 2.5 metre depth of water. The length of pond may be extended as per convenience, but the width should be limited to 40 metres. Fully controlled inlets and outlets should be provided to regulate the inflow and outflow of water in the pond.

Soil Type

Next to topography, the soil type is an important factor to be considered. The soil should be free from vertical and lateral seepage. As such, the soil with limestone, sand, rocky bottom, gravel deposit should be discarded. Porous soil is the indication of a poor condition of soil because, from such soil, the chance of seepage is always open. Ideal soils are impervious, ordinarily heavy, silty clay, clay loam etc. Alluvial soil is the best for a successful fish culture. In case the pond has porous soil all-around, the bottom of the pond can be given treatment of betonite clay or other soil sealants, as per expert advice (Pruginin and Ben Ari, 1970; Banerjee and Thakurta, 1979). Care should be taken for strengthening the embankment by scarification and disking efforts.

Water Supply

Another point of consideration for construction of a pond is the prospect of availability of an adequate and independent source of water. The source may be dam, canal, lakes, reservoir, river, stream, spring, etc. The surface run off, artesian wells, tube wells may also serve the purpose. However, in respect to regular water supply, if the pond has elevated location, reservoirs and springs are considered as best sources, being dependable throughout the year. Streams or canals are also dependable sources as the water flow is adequate enough in volume as well as maintenance of level to provide constant flow. Above all these, the silt flow is the least.

An ideal suggestion is that the water should enter the pond above 8′ from the water level of the pond and an inlet and an outlet must be provided to a pond. Regular supply of water to maintain proper level in the pond throughout the year enhances the productivity prospectives. However, care should be taken to see that the supply water system should not come in contact with any type of pollutant through inter-connected drain, channel or lane.

Construction of Pond

Usually the size of the farm and its location are two important factors in selecting a fish farm site. The size depends on the purpose of the farm. If the subsistence fishing only is the aim, small sites (not smaller than 0.4 ha) may be used. If the aim is of multi-purpose kind including commercial and experimental, the fish farm may comprise nursery, rearing, stocking, breeding ponds units for which a larger area will be required. In India usually four kinds of ponds are needed for fish culture (Fig. 8.1).

Hatching pond. It is a small pond meant for hatching of eggs. Its usual size is 20′ × 8′ × 2′. It gets provision of hapa, the size of which is generally 6′ × 4′ × 2′. As the water level is low in the hatching ponds, it dries up, in the summer.

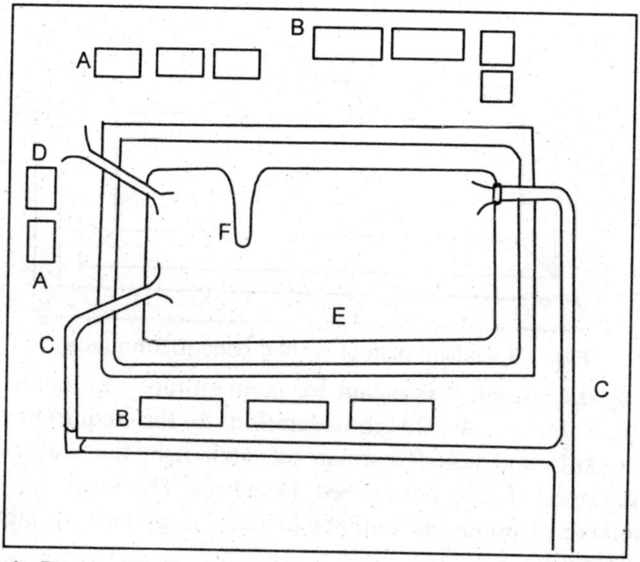

Fig. 8.1. Design (Outline) of a Fish Farm : Nursery pond (A), Rearing tank (B), overflow channel (C), water inlet (D), main tank (E), feeder tank (F).

Nursing pond. It should preferably be small (50′ × 50′ × 4′) shallow with 1.0 to 1.5 metre depth of water (Fig. 8.2).

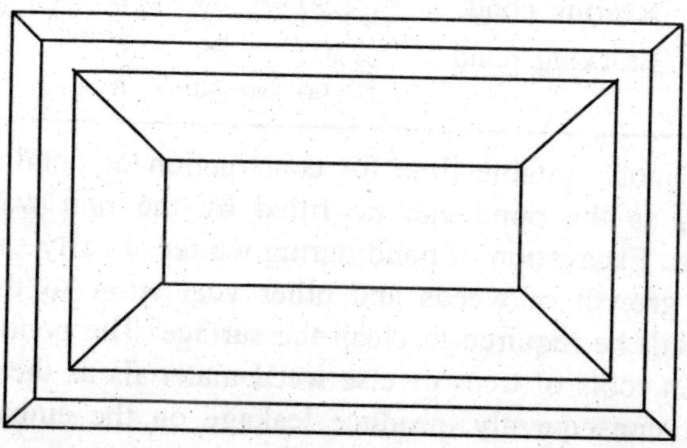

Fig. 8.2. Outline plan of a standard nursery pond (0.02 ha area).

Rearing pond. It should be slightly larger than the nursery pond but need not be deep. It should have 1.5 to 2.0 metre depth of water. An ideal size of the rearing pond is 90′ × 30′ × 4′ (Fig. 8.3).

Fig. 8.3. Outline plan of rearing pond (0.1 ha area).

Stocking pond. It is meant for pond culture and its ideal size is 300′ × 80′ × 60′. In consideration to the requirements of smaller units and less fisherman to catch fish, the width of the stocking pond should not exceed 40 metres. The ideal suggestion for the size of pond and number of units is given in a tabulated form below :

Sl. No.	Type of ponds	Size of ponds (feet)	No. of units
1.	Hatcheries	8 × 4 × 2	6
2.	Nursery pond	50 × 50 × 6	4
3.	Rearing pond	90 × 30 × 6	2
4.	Stocking pond	300 × 90 × 6 90 × 30 × 6	6

The most suitable time for construction of pond is the late summer, as the pond can be filled by the rain water during monsoon. Excavation of pond during winter or early summer will lead to growth of weeds and other vegetation so that manual labour will be required to clean the surface. The pond should be free from roots of trees or else weed materials as they will later rot and, consequently, produce leakage on the embankment at about 5 metres. From the pond water line, trees and bushes should be cut and removed so that sunshine may be reaching the water area of the pond and dried leaves, flowers, seeds or fruits may not fall in water.

Keeping in view the drawing or design prepared by the experts, the construction work is taken up in the form of excavation by manual labour. The bottom area and sides of the pond should be rammed to a proper level as per design. When the digging work is completed, it is necessary to remove the top soil within the outlined area by about 0.3 metres as the top soil generally contains roots of shrubs and trees and decaying organic matters which can deteriorate the quality of water and provide hindrance to fish culture practices. Soon after removal of the top soil, the core tench is cut. To prevent the excessive seepage, clay core should be provided as mentioned before also. Earthen embankment alaround the pond should be made by filling it with earth blocks, obtained from the earth digging.

Fig. 8.4A. Outline plan of a stock pond (0.4 ha area).

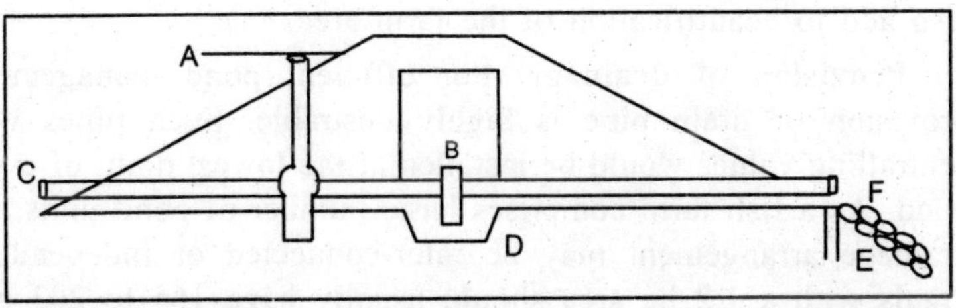

Fig. 8.4B. Cross section view of the dam area of a fish pond : A water line (A), Core (B), Drain pipe (C), clay-footing (D), Concrete footing (E), Stone Rubble (F).

Embankments. The slope of embankment should be constructed in a scientific manner, *i.e.*, on giving consideration to the distance in horizontal axis for each part of height (Fig. 8.4A) A 2.1 slope means 61.0 cm of base for 30.5 cm of height. Quality of soil is a factor to be considered for this construction. If the soil is good clay, this slope can be 2:1 on the inner side of embankment, *i.e.*, towards water side, but it should be 3:1 for loamy, silly or sandy soils. According to Marr *et al.* (1966), the crest of embankment should not be less than 91.5 cm wide (Fig. 8.4B). If the pond is constructed in swampy areas, the crest of embankment should be at least 122-152.5 cm wide.

All earth embankments should have some extra height above the water level to prevent waves and flood from overflowing and washing them out. The extra height is called free board. For ponds having 0.4 to 1.2 ha water area, the free board should be 60 cm, and in larger water bodies, free board may be up to 90 cm. In heavy rainfall areas, a free board in 60 cm extra height is necessary for nursery and rearing ponds, but if the region has low average rainfall a 30 cm free board would serve the purpose. Expert suggestion is that at the time of construction, the embankment should be 10% higher than present requirement to adjust with the settlement of the earth in due course of time (Delmendo *et al.*, 1970; Saha and Gopalkrishnan, 1974).

As soon as embankment is completed, turfing with low creeping grasses with short stolons should be done to prevent erosion. A fencing with bamboo or barbed wire is needed around the fish farm so as to keep the cattle off the embankment. Raising multiflora rose around embankment may provide good fence and also add to beautification of the farm area.

Provision of drainage. For efficient pond management, provision of drain pipe is highly desirable. Such pipes with centralling values would be installed at the lowest point of every pond. As a fish farm comprises large number of pond units, the drainage arrangement may be inter-connected or independent. Ponds with a 1.2 ha area should usually have 165 to 203 mm diametre drain pipe, while for large water areas, 203 to 305 mm pipe is necessary. In fact the size of the pipe depends on the size of the pond and the volume of water that will be passing through the pipe. Asbestos cement pipes, concrete spun pipes, east iron pipes and galvanized fibre glass pipes, terracota concrete tiles are commonly used as drain pipes. However, the best material is asbestos cement pipes for a large pond and cast iron pipes and glavanized iron pipes for small ponds.

Inlet pipes. It should be at least 155 mm above the water level of pond, when the latter is full and they should be provided with a suitable screen so as to prevent the entrance chance of wild fishes into the pond, as well as the escape possibility of the pond fish. Pond inlets may be of simple furrow type and those made of brick or concrete with slots for sliding a strongly built control board and screen.

Mechanical spillway. In a successful functioning of stock ponds, a mechanised spillway is a necessity as it controls all the normal overflows and acts as a safety outlet to prevent ponds from flooding during heavy rain.

After the construction work, filling of the pond with water is the next necessity. During the initial filling a great care is needed to avoid slumping and sliding of the wall due to forceful water flow. Draining should be done slow so that the embankment may not get damaged.

Another aspect of consideration is related to the creation of waves due to speedy winds. Large waves may severely erode the face of the embankment above and below the line. To catch this, a simple long wall should be built along the face of the embankment. This should break the waves or weaken it and, thus, help to check chance of erosion of the wall.

Sometimes embankments get damaged due to burrowing by rats and crabs. The best device to control this is to trap the crabs or rats or plug these burrows with compacted cinders. The use of suitable poison may help to control the entry of crabs and rats.

In some cases, the ponds may have muddy water so that the water turns turbid, as the clay particles remain in suspension indefinitely. To get rid of this situation, the use of agriculture gypsum (hydrated calcium phosphate) is most helpful. The recommended rate is 192 kg per 1000 m^3 of volume of water. On scattering the chemical evenly over the entire pond surface, it has been found that the clay particles come together and then settle at the bottom. The water consequently turns clear. If the effect is not adequate, addition of about ¼ of the amount to the above suggested quantum in 6 weeks time may cause control of turbidity effectively.

Maintenance of Fish Pond

Before stocking of the pond, several measures are to taken so as to get better result and to have a successful fish culture. Control of weeds, and also of fish enemies (*e.g.* predaceous organisms like aquatic insects and carnivorous fishes), provision of liming and fertilizers are the main requisites. Detailed description of these requisites are given ahead in the exclusive chapter 'Pond Fertilization'

9

Quality of Water Media and Bottom Soil

Both water and bottom soil are directly concerned in the culture practices of fish. It is now a known fact that about three-fourth of the earth's surface (that is, about 71% surface area) is covered with water. The latter represents the most extensive living media for animal and plant life. Also water is a biological necessity for all kinds of animal and plant life.

The most suitable soil for pond construction is clay loamy soil, mart and peat are also satisfactory, but the sandy soils are not suitable at all because their water holding capacity is very low. The depth of water in the pond is kept between 2-3 metres and in no case less than 1.5 metres. The average silt should be 20-25 cm uniformly spread over bottom.

The bottom soil of the pond provides a base to the water media, releases nutrients and induces for manifestation of the planktonic and other organisms. The chemical and physiological process of organisms involve utilization of water in some form or the other. As such, one can say that water plays an essential role in several life activities.

Water is the basic element in fish culture and its specific properties as a cultural media are naturally of great significance in the productivity of pond. Pure water cannot support living organisms, but its containment of nitrogen, phosphorus, potassium and calcium salts, dissolved organic matter and the gases like oxygen, nitrogen, and carbon dioxide determine to a large extent the activities of aquatic biota.

The physico-chemical properties of pond soil are of greater significance than is generally realised. When soil conditions are not favourable, the production potentiality is directly affected. Therefore, during construction of a new pond, first and foremost consideration is of the soil; the water supply question arises after the construction of the pond.

Unlike the marine environment, which is known for its constancy of conditions, the freshwater environment is subject to variations in the environmental factors, such as temperature, dissolved gases, turbidity, light and food organisms. The freshwater is characterised distinguishly by low salinity and, hence, low osmotic pressure. The salt content, on average, is 1.8 per cent *i.e.*, 18 parts per thousand. During rainy season, there are 30 to 40 parts of solutes per million parts as the solutes are swept from different sites and they reach low lying areas including ponds. In the freshwater bodies, salts are present in meagre quantity, mostly in the form of carbohydrates whereas in the sea the salts (mostly sodium chloride) are contained in much higher percentage, *i.e.*, on average, 3.5 per cent. Some salt lakes may have the salinity as much as 25 per cent to 30 per cent. This status of the lake in question greatly restricts life in it.

Status of Cultivable Water

Some basic characters of the cultivable water are given below. In shallow ponds, question of thermal stratification does not arise or if it is there during rainy reason, when water is full, the stratification is not prominent. In tropical ponds, a daily turn over of water bodies takes place. During night there is more circulation of water so that there is mixing of the water contained in the enclosed water body and there is circulation of oxygen and nutrients. Experimental results have shown that the life processes get enhanced by an increase in temperature. The resultant effect is on the level of dissolved oxygen which, after more consumption, gets decreased proportionally.

Turbidity. It has an important role to play in the productivity of a pond. The transparency of water, *i.e.*, the least level of turbidity in water enables the penetration of light to the depths where it is ultimately absorbed. In deep water or reservoir water, the zone upto which the light rays penetrate is called "photic zone" and below this zone, there is complete darkness and the organisms that require light (*e.g.* algae) cannot live. Sometimes manifestation of phytoplankton due to the application of organic manures in excess adversely affects the productivity of pond. Photosynthetic activities of plants in turbid water get reduced and the fish fauna needing clear oxygen-rich water are replaced by predominately labyrinthiform fauna in such a situation.

Dissolved oxygen. One of the most critical factor in the aquatic environment is the amount of oxygen in water, as all fishes as well as their food organisms depend on dissolved oxygen for their respiration. The aerobic bacteria present in water are also major consumers of dissolved oxygen. Besides the atmospheric oxygen that get dissolved in water by diffusion on its surface, plants release oxygen during the process of photosynthesis. The water circulation in the tropical ponds helps mixing of the dissolved oxygen so that, on average, a concentration of 5 ppm at a temperature of over 20°C is maintained and this is consumed by almost all organisms (except anaerobs, *e.g.* anaerobic bacteria).

Biological oxygen demand. The biological oxygen demand (BOD) is the used up oxygen by unstable organic matter due to bacterial action. It is an indication of amount of oxygen present in a pond. Any rise in BOD value results in depletion of oxygen and, in critical level of occurrence, mortality of fish may occur. In most of rural pond waters, with organic deposits at the water bottom, BOD of the order of 2 ppm may occur.

Hydrogen-ion-concentration (pH). The pH of a water media oftenly gives indication of fertility or potential productivity in a pond. A slight alkaline medium (*i.e.,* pH of 7 to 8) is considered as ideal, being suitable for fish life. This pH condition is of great help in the conversion of the organic matter into assimilable substances, such as ammonia and nitrates which are nutrients for increasing productivity in pond. On the other hand, acidic condition of pond (below 7 pH) adversely affects the life of aquatic biota, as impeding in the circulation of nutrients takes place and also inhibition in the nitrogen fixation occurs. Remedy at this condition of water lies in the application of lime with calcium carbonate, *i.e.*, alkaline fertilizers.

Carbon dioxide. The decomposition of organic matter and the breathing activities of aquatic plants and animals produce carbon dioxide. This gas is one of the essential raw materials necessary for photosynthetic activity by the green plants. A high level of carbon dioxide content of water is lethal to the fish life and it prevents the oxygenation of water. Nevertheless, it may also affect the extraction of dissolved oxygen by living organisms from the water. A lethal limit of this gas for healthy fish life is 5 ml per litre of water.

Carbon dioxide combines chemically with water to produce carbonic acid (H_2CO_3) which influences the pH level of water. The carbonic acid dissociates to produce hydrogen (H^+) and bicarbonate ions (HCO_3^-) as per following equation :

$$CO_2 + H_2O = H_2CO_3 = H^+ + HCO_3-$$

The amount of free or uncombined carbon dioxide in water is of ecological importance.

Nitrogen. Nitrogen is one of the basic elements of fertility in ponds and is present in water in an amount of 65 per cent of all dissolved gases at equilibrium. Certain quantities of nitrogen may reach soil and water from the atmospheric nitrogen by rain or incidental lightening. There is also role of soil bacteria that can fix atmospheric nitrogen and, thus, make it available to the green plants. Whatsoever are the alternative possibilities, the ponds have the major supply of nitrogen from the decaying organic matters, *i.e.*, from the putrefaction cycle in which nitrates are the end-products. From the productivity of ponds, sufficient quantities of nitrogen compounds, such as nitrates are essentialities when water contains 4 ppm of nitrogen with 1 ppm of phosphorous and 1 ppm of potassium the production of plankton is high.

Hydrogen sulphide. The deeper layer of many ponds, tanks, lakes, reservoirs and some estuaries may contain significant amounts of the toxic gas, hydrogen sulphides, which is formed by decaying organic matters. In this decaying action, anaerobic bacteria plays the main role. Sometimes the gas accumulates in the fish ponds having thick layer of organic deposits at the bottom. This gas can be a severe limiting factor in fish ponds, as a concentration of 6 ppm of H_2S will yield drastic effects for the acquatic biota in general and can kill common carp in few hours. In such a situation, aeration of pond by mechanical agitation can help a lot to free the pond of hydrogen sulphide.

Phosphorus. Phosphates and nitrates are essential for fish production in ponds. In all natural waters, these nutrients are found in varying quantities in dissolved form in the bottom soil. If nutrients are deficient, production and growth of organisms in the pond water are adversely affected. The phosphorus is released from the dead organisms in the bottom of the pond of giving rise to the growth of 1 ppm of phosphorus. This level of phosphorus is considered as optimum for sufficient growth of plankton.

Potassium. This is present in a cultivable pond to an extent of 1 ppm along with 1 ppm phosphorus and 4 ppm nitrogen. This concentration of the three constituents determine the fertility of the pond and is considered to be most ideal for the growth of planktons. Potassium is released from the bottom soil which is rich in deposits of various kinds. It is released from there and is specially effective in stimulating the growth of aquatic vegetation (Matida, 1967).

Interrelationship Between Bottom Soil and Water

Water fertility contributes largely to the production of plankton but soil fertility is mainly effective in the growth of benthic vegetation. At the pond bottom there is the containment of organic deposits and mineral substances. These produce the essential nutrients which, by chemical or bio-chemical actions and interactions increase the pondwater supply. Nutrient materials are released into the water and these help in the fixation or chemical binding of such released substances or of those introduced in the form of fertilizers or manures. All these help in the production of planktons and green vegetations. It is an admitted fact that production in ponds with a bottom rich in organic deposits and minerals is much greater than in those with poor soil. The colloidal substances of the top muddy layer is of importance in respect to enrichment of the pond water for production capacity.

To retain the production capacity of the pond, a full drainage of used water is done before rainy season and then replacement with freshwater is carried over. When the pond is dried up in the summer after drainage, *i.e.*, after harvesting of fish crop followed by drainage, the bottom soil is directly exposed to the sunlight and atmospheric air. These factors help to kill the parasites and other disease carrier organisms. At this time, planting of leguminous plants, such as pulses plants, help to enrich the nitrate substance and increase the bacterial growth in the soil. This aspect has already been dealt with before in the chapter 'Fish Farm Construction'.

Impact of Pond Fertilization

Continued culture practices and harvesting for some years in the fish ponds lead to removal of large amounts of nutrient substances out of the pond ecosystem. In this situation, it is a call of the time that for getting a good return from the pond in future, there should be proper management for provision of fertilizers or organic manures from external sources, both organic and inorganic fertilizers (Saha, 1969; Lin, 1970; Chakrabarty *et al.*, 1975).

It is advisable to endeavour for preparing the schedule of manuring after examining the quality of bottom soil. If the bottom soil is highly clayey with rich organic deposits in the top soil, inorganic fertilizers with lesser amount of organic manure give good results. If the bottom soil is neither too clayey nor sandy, *i.e.*, is sandy-clayey but with organic deposits to a considerable level, a combination of both organic manure and inorganic fertilizers is recommendable. However, if the pond has acidic bottom soil (*i.e.*, with low pH), these should be pretreated (a weak before) with quick lime (CaO) to get beneficial results.

This advice for undertaking liming action is in consideration to many reasons. Besides neutralizing the acidity, it precipitates colloidal matter, already existent in the water, promotes decomposition of the organic matter through bacterial actions and, nevertheless, provides calcium to growing plants and fishes for bone formation. The role of calcium, in collaboration with carbon dioxide, in preventing wide diurnal changes of pH of water, is most remarkable.

Based on degree of acidity or neutrality or alkalinity, lime dose schedule should be like this.

Soil type	Soil pH	Lime dose (kg/ha)
Highly acidic	4.0 – 5.0	2,000
Moderately acidic	5.0 – 6.5	1,000
Almost neutral	6.5 – 7.5	500
Mildly alkaline	7.5 – 8.5	200

There are other ecological factors to be considered before fixing up the schedule. It would be better to consult some experts of the agricultural universities (I.C.A.R. Undertaking) or other experts of the state or central government for an authentic suggestion.

Basic Considerations in Nursery Pond Management

A nursery pond is generally rectangular in shape, has a surface area of 200-400 m^2 each. However, any small seasonal pond with shallow water can serve the purpose. A nursery pond constructed in a planned manner should have the features as given below :

1. The physical texture of the pond soil should be loamy or clay-loam so that the power of retention of water is good.

2. The bottom of pond should not have any amount of decomposing organic matters.

3. The physico-chemical features of the pond soil and pond water should be typically conducive for rich planktonic production. As such, the bottom soil should have these features :

Soil pH : 6.5 – 7.5; available nitrogen (N) : 30-50 mg/100 kg, available phosphorus : 6.0-16.1mg/100 g and organic carbon : 1.0-2.0%.

On the other hand, the water contained in the nursery pond should have turbidity : below 20 ppm, pH : 7.0-8.5, total alkalinity : 75-150 ppm and dissolved oxygen : 4-10 ppm, free CO_2 : nil – 77 ppm, bicarbonate : nil – 250 ppm, total alkalinity : 50-280 ppm, nitrate : 0.05-1.5 ppm and phosphate : 0.05-7.0 ppm.

4. Abundance of vegetation is not desirable in nurseries, because there would be wastage of soil nutrition, arresting of

planktonic productions, obstruction to free movement of fish fry in the pond water and hindrance in undertaking netting operation. Water should be available for at least 3 months to maintain a steady water level between 0.75-1.0 metre. Nursery pond with alkaline water and fair level of phosphate and nitrate can, in a normal situation, yield satisfactory growth and the survival chance of fry may be on a higher side.

5. The tiny fish fry in the nursery pond may have enemies like all species of fishes in large size, particularly carnivores and also frogs, snakes and birds. All unwanted fishes like mussels, catfishes bring havoc in the form of destruction to hatchling and advanced fry; their complete eradication is another important necessity before stocking nurseries.

Supply of Water

Water from any source, such as spring, river, canals, surface run-off from rainfall can be used for refilling nursery ponds, provided the water was not contaminated and undesirable substances were not contained by it. To be on safer side, it is desirable that the water should be passed through a fine sieve to get rid off predators and insects in any stage of life cycle. Generally nursery ponds are filled with water 10-15 days before stocking, initially to 60 cm depth, but later the depth can be raised to 100-120 cm.

Pond Fertilization

In case stocking is done in high densities, the available food reserve of the pond may be used up in a very short time and scarcity of food may arise. Giving weightage to this possibility, it is advisable that manuring should be continued in such a regularity that planktonic growth is maintained for a larger period (Chakrabarty *et al.*, 1975; Sinha and Ramachandran, 1985). The ideal procedure for pond fertilization is like this :

Lime should be supplied after two weeks of eradication of predatory and weed fishes. This helps in correcting the acid condition of soil and water acidity, *i.e.*, for creating a strong pH buffer system, decomposition of organic matter, and also for serving as disinfectant against fish diseases, participating in the preparation of harmful metallic compounds, providing chance of aeration for improvement of humus and, nevertheless, for manifestation of planktons (phyto- and zoo-plankton) as the calcium acts as nutritive substance.

The above functional capacities of lime ultimately induce planktonic growth. The carp spawn grow fast by consuming planktons (zooplanktons in dominance over phytoplanktons). Fertility of pond is increased by the application of inorganic and organic fertilizers as a combination of bottom life activities of the pond depends on the condition of the bottom soil regarding bottom deposits and minerals and on water quality. The fertility can be enhanced by applying fertilizers or manures which, by decomposition and re-combination provide required nutrients, minerals, vitamins to the water media and consequently, a high production of planktonic biomass occurs. In this connection, a recent information has been on record that nitrogen, phosphorus, potassium and calcium are directly concerned in increasing the pond fertility. However, trace elements like cobalt, manganese, iron, boron, copper and zinc play important role in this consequence, that is, increase of fertility (Banerjee and Banerjee, 1967).

BASIC CONSIDERATIONS IN REARING POND MANAGEMENT

Rearing ponds are smaller than stocking pond, but larger than nursery pond. The size of the rearing pond is generally 0.08 to 0.1 hectare in area and the shape is rectangular. The depth of water ranges from 1.2 to 1.6 m and seldom more than this. The management method and practices *e.g.* technique of pond preparation, pond cleaning and fertilizations are practically the same for the rearing pond as described before for nurseries. The fries, generally of 20-25 mm size, are reared in the rearing pond to 100-150 mm size, normally for a period of 2-3 months and, in exceptional cases, upto 6 months. It needs precise mention here that rearing of fry to fingerling size in a separate pond (*i.e.*, rearing pond) is necessary, as the small fry in the nurseries have choice of planktonic food for consumption when compared to the condition of stocking pond (Das, 1967). Secondarily, the presence of predatory fishes in the larger pond (*i.e.*, stocking pond) would adversely affect the stock position of fish and prospect of harvest in the stock pond would be drastically affected. In the rearing pond it is essential to add manure first so that fertility of the pond would increase and this would induce planktonic growth.

BASIC CONSIDERATIONS IN STOCK-POND MANAGEMENT

The most suitable soil for pond construction is clay and the sandy soils are the worst as their water holding capacity is very low. Loamy soil, marl and peat are also recommendable after preference over clayey soil. Rural ponds, which are usually not constructed as per any scientific plan and lack certain basic requirement such as lack of drainage system, having no perfect enbankment, and so on so forth, yet, they carry potential value for fish culture. With some expenditure on construction improvement, one can make use of these ponds for stocking fingerling, rather than getting involvement in huge investment to construct new ponds (Ganapati and Chacko, 1950; Jingran, *et al.*, 1979).

In addition to these village ponds for use in fish culture, new ponds can be constructed as per scientifically known plan and fingerlings from the rearing ponds can be transferred there for picking up growth to marketable size. The new constructed ponds are generally rectangular in shape, have gradual sloping banks, even bottom and firm embankments. They are provided with gradual inlet and outlet. The size of the pond is variable, generally in a range of 0.2 ha to 0.5 ha. In experimental cases, size can go up to 2 ha in order to have bigger stocking capacity. The depth of water in the pond should be maintained to 2.3 metres, but in no case below 1.5 m, as the sunlight (*i.e.*, temperature effect) would be reaching the bottom and the fingerlings would be in uncomfortable situation in several respects. The silt which accumulates along with supply of water, should be, on average, between 20-25 cm, uniformly spread over the bottom.

Before introducing fingerling in the stocking pond, due attention in the pre-stockage operation is needed. Water weeds have to be removed and piscivorous predators like the freshwater shark (*Wallago attu*), murrels (*Channa* species), and also weed fishes like carp minnows (which compete with culture fishes for food consumption) should be removed by netting (for predators) or chemical treatments.

The survival chance of the fingerling can be almost absolute, if proper care has been taken as pointed out before. Obviously proper management can give high return of fish yields at the time of harvesting and whole investment on the management would prove economical.

In the stock-pond, the application of organic manures should be given 2-3 weeks before stocking fingerlings. This is followed by manuring in a number of instalments during the post stocking period.

The application of organic manure should not be done in an ordinary way, *i.e.*, by broadcasting over the surface of water; rather, the proper method to be followed is that it should be put first in baskets and then kept under water in the corners of the pond.

The schedule of manuring should be adjusted for application as per following considerations :

1. Estimation of available food reserve in the pond (which is based on assessment of plankton both qualitatively and quantitatively).

2. Estimation of nutrient level in the pond which is related to quality of soil and water.

3. Observation of physico-chemical condition of the water and climatic condition.

4. Observation of average growth rate of the fish stocked in the pond.

BASIC CONSIDERATIONS IN SEWAGE FED PONDS

The use of sewage effluent for raising fish yields from ponds has been recognized comparatively late in India. Several fishery scientists observed, by experimental works, great potentiality of sewage fed ponds for fish culture. Later, Saha *et al.* (1958) and Ghosh *et al.* (1976) reported physico-chemical properties of the Calcutta sewage and of the ponds fed by the sewage in diluted form in 1:2 ratio.

The physico-chemical features, of domestic sewage (drawn from a sewage plant of Kolkata and of pond water located near Barrackpore (West Bengal) are given here :

Parameter	Pond water*	Sewage Effluents
Temperature °C	27.8 — 31.00	27.5 — 30.8
pH	8.1 — 8.2	6.9 — 7.2
Dissolved Oxygen	2.9 — 6.4	Nil-Traces
Carbon dioxide	Nil — 10	52.0 — 84.0
Alkalinity	196.0 — 266.0	224.0 — 516.0
Ammonical nitrogen	0.16 — 5.8	3.2 — 24.0
Nitrite-N	0.16 — 0.4	Nil
Nitrate-N	Nil — 0.08	0.14 — 0.5
Albuminoid-N	Nil	0.11 — 2.35
Phosphate	0.5 — 6.0	3.6 — 12.6
Chloride	82.0 — 100.00	45.0 — 101.5
Calcium	36.0 — 31.0	66.2 — 80.0
Magnesium	14.7 — 25.2	16.4 — 35.3

* After dilution with the sewage effluent in the ratio of 1:2.

Fishes like magur (*Clarias batrachus*), tilapia (*Tilapia mossambica*) can be reared in such sewage fed ponds and high yield can be obtained (Krishnamoorthy *et al.*, 1976; Ghosh *et al.*, 1979). Also some major carps, *e.g.* mrigal, which have the habit of consuming decaying organic matters from the bottom, are purposely cultured in sewage fed pond water and high yields are harvested.

The significant part in the composition of sewage effluents is that on decomposition sewage shows the containment of high percentage of nitrogen, phophorus, calcium, potassium, etc. All these elements have nutritive values and these contribute largely to a healthy productivity of the sewage mixed water.

Gopalkrishnan and Srinath (1963) have recommended the use of treated domestic sewage also known as 'activated sludge' instead of diluted sewage effluent in consideration to its more effectivity, easy transportation in the dried form and possibility of its storage for use whenever demand arose. From the results of laboratory experiments it has been inferred that the sludge has a direct influencing effect on the production of plankton, and on the growth of fish. The obvious reason for this positive effect is that there is direct release of the nutrients, contained in the sludge, into the pond waters.

The plankton which flourish in pond water fed with sewage effluents are several but the dominating types are the following :

Zooplanktons. Daphnia sp., Monia sp., Cyclops sp., *Brachionus* sp., *Keratella* sp. and Copepods sp.

BASIC CONSIDERATIONS IN BRACKISH WATER PONDS

The practice of fish culture in brackish water has gained popularity recently for getting yields of selected fish, *i.e.*, milk fish and of prawn. The ponds are constructed at such a site near the sea or adjacent to the mouth of estuaries that salt water is received at every high tide. A tidal range of 1.5 m is considered essential for a brackish water fish farm.

Preferably the area of the pond is 0.4-2.4 hectare, and depth of water is 0.3 to 1.3 m. There is a well built bamboo or wooden sluice which controls the ingress and egress of water. The bottom of the pond is of clayey soil which is rich in organic deposits and minerals. The embankment also is usually constructed with clayey soil. One gross difference between fresh water pond and brackish water pond is that the latter is self-sufficient farm with nursery or rearing and stock ponds units in the same enclosure.

In Philippine, the size of the farm is generally 10 ha. There is arrangement of 1 per cent of the pond area for nurseries, 9 per cent for rearing pond and 90 per cent for stocking ponds. The height of the embankment of the brackish water pond in India is determined by the height of the tide. Generally 30 cm height serves the purpose well. During construction, attention is given to keep the slope of the pond water towards the outlet.

The clayey bottom soil of the pond keeps the brackish water highly productive. It has been observed that the incoming tides from seawater bring organic matters which spread in the column of water and gradually get deposited at the bottom soil. This system adds to the fertility of the soil. Usually there is growth of some common species of plants like Azobacter, Nostoc, Anabaena. These plants fix atmospheric nitrogen and thus nutrient content of the soil increases. They help in increasing the fertility of the water and growth of planktons. The water of the brackish water pond is always alkaline with the overall effect that there is little variation in pH. The oxygen content is subjected to variations in temperature and salinity, but the variation margin in the oxygen level is within limits. There is seldom occurrence of depletion in oxygen content.

The variation is salinity is noticeable due to two sources of water infeeding in these ponds : in daily incoming of tidal water from the estuaries and rain water during monsoon period. Daily supply of the tidal water from the estuaries causes daily fluctuation in the salinity level. However, during monsoon, period large quantity of freshwater is brought to, the pond by rains so that pond water gets diluted and salinity level decreases. Reverse to this, during summer there is loss of water and, in consequence to this, salinity increases. These sorts of fluctuations in the water quality, particularly in the salinity level, have selective effect. As to fishes, the effect is more distinct, and only euryhaline species are in a position to thrive.

Among algae, several genera and species of *Myxophyceae*, *Diatomeceae*, *Chlorophyceae* grow abundantly, and among crustaceans the cladocerans, ostracods and copepods have luxurious life. A remarkable fact is that the consumer of zooplanktons are conspicuously less.

The general method followed in the preparation of brackish water pond for stocking is almost the same as done in the case of fresh water pond. Here also the fertility of the pond can be increased by draining and drying pond bottom.

The manuring is required to stimulate the growth of phytoplankton, such as *Myxophyceae*. For accomplishment of this purpose, green manuring is done in Java soon after draining, prior to stocking. Transplanting algae with the addition of some artificial mutrient is practised in Philippines. In Hong Kong, there is manuring of the pond with the rice bran, pea nut-cake, pig manure and night soil. After completing manuring action water is supplied in the pond and the latter is left as such for about two weeks. After this, the pond is stocked with fingerlings.

All these informations are indication of the fact that the pond water and bottom soil have great role to play in enhancing the potentiality of the pond for fish food and fish production (Devraj and Seenappa, 1992). All efforts to raise fish production from a pond may get defeated if quality of water of the pond and of the bottom soil is not at satisfactory level.

REFERENCES

Banerjee, S.M. 1967. Water quality and soil condition of fish ponds in some states of India in relation to fish produciton. Indian J. Fish 14 (1 & 2) : 115-44.

Banerjee, S.M. and Ghosh, S.R. 1970. Studies on the correlation between soil reaction and different forms of bound phosphorus in pond soils. J. Inland Fish Soc. India 2 : 113-20.

Banerjee, S.M. 1971. Soil study in composite fish culture. First workshop on All India Co-ordinated Research Project on composite fish culture. ICAR Cuttack, September 1971.

Banerjee, S.C. and Thakurta, S.C. 1979. Effects of physico-chemical methods in controlling seepage in ponds (Abs.) In : Symp. On Inland Aquaculture, February 12-14, 1979 : 17 CIFRI, Barrackpore.

Chakrabarty, R.D., Sen, P.R., Chatterjee, D.K. and Jena, S. 1975. On the use of fertilizer and supplementary feed for enhancing fish production in freshwater ponds. Proc. Nat. Acad. Sci. B (45) 3 : 192-96.

Chatterjee, D.K. 1979. Effect of lime on fish pond soils. In Symposium on inland Aquaculture, Feb. 12-14, 1979.

Das, B.C. 1967. Effects of micronutrients on the survival and growth of Indian carp fry. FAO Fish Rep 44 (3) : 241-56.

Delmendo, M.N., Osborn, P.E., Halili, R. and Martinex, E. 1970. Construction of ponds for aquaculture. Proc. Indo-Pacif. Fish Coun. 14th Session, IPFC/C 70/SYM 26 : 30 p (mimeo.).

Edminister, F.C. 1947. Fish Ponds for the Farm. New York : Charles Scribner's Sons, 114 p.

Ganapati, P.N. and Chacko, P.I. 1950. Suggestions for stocking fish ponds in Madras. Madras agric. J., 37 : 1-5.

Ghosh, S.R., Rao, N.G.S. and Mohanty, A.N. 1979. Studies on the relative efficiency of organic manures and inorganic fertilizer in plankton production and its relation with water quality. In : Symposium on Inland Aquaculture (Abstract), February 12-14, 1979 : 107-08 CIFRI : Barrackpore.

—— and Srinath, E.G. 1963. Some experimental observations on the use of activated sludge for fish culture. Proc. Ind. Acad. Sci. 57B (6) : 379-88.

Hora, S.L. and Pillay, T.V.R. 1962. Handbook on fish culture in the Indo-Pacific region. FAO Fish Biol. Tech. Pap. (14) : 204 p.

Lawrence, J.M. 1949. Construction of farm fish ponds. Cire. agric. exp. sta. Albama Polytechnic Inst. Auburn (3) : 55 p.

Lin, S.Y. The effect of phosphorus and nitrogen fertilizers on fish production in fresh-water fish ponds by the use of inorganic fertilizers. FAO Fish. Rep. (44) 3 : 210-25.

Matida, Y. 1967. The role of soil in fish pond productivity in Asia and Far East. FAO Fish Rep. (44) 3 : 54-63.

Pruginin, J. and Ben Ari, Aza. 1959. Instructions for the construction and repair of fish ponds. Bamidgeh 11 (1) : 25-28.

Rounseefell, George A. 1985. Fishery Science. International Books and Periodical Supply Service. 24B/5 Deshbandhu Gupta Road, Karol Bagh, New Delhi-5.

Saha, G.N. 1969. Relative efficiency of different nitrogenous fertilizers in fish pond soils. J. Indian Soc. Soil Sci. 17 (1) : 51-57.

Saha, C. and Gopalkrishnan, V. 1974. Construction of fish ponds. J. Inland Fish Soc. India 6 : 122-30.

Sinha, V.R.P. 1975. Freshwater fish farming for more income. Indian Fng. 28 : 107-08.

—— and Ramachandran, V. 1985. Freshwater Fish Culture, Indian Council of Agricultural Research, New Delhi.

Worbel, S. 1967. The role of soils in fish production in ponds. FAO Fish Rep. (44) 3 : 153-63.

10

Nursery Pond Management : Fry Production

With proper application of fish culture management technique and bringing all fish water resources in the fold of scientific basic of field culture, substantial increase in fish production through the fresh water fish culture practices could be anticipated. At present 44.77 million hectares of water area is under fish culture practice out of which 28.55 million hectares are already available in India as ponds and tanks. Unfortunately India produces 1.3 million tonnes of fish from fresh water sources, mostly from fish farms. As such more incentive is needed towards expanding areas of cultivable ponds and tanks and standardising culture practices through the scientific management of nursery pond for fry production, rearing pond management for fingerlings production and stock pond management for harvesting the fully grown (*i.e.*, adult) fish for domestic consumption and marketing and, thus, providing substantial revenue source to the country. Per capita consumption of fish has been 3.2 kg on an average upto 1992 (5.13 kg for fish eating population/yr against the estimated requirement of 11.0 kg).

The site of nursery pond or tank is generally close to the spawn collecting grounds so as to facilitate the release of freshly collected fry directly into them. The ideal size of the nursery pond is a square of 50′ × 50′. However, a pond in rectangular shape with an area of 200-400 m^2 is also recommended in some areas (Fig. 8.2). The modification can be to such an extent that a nursery pond can have cemented structure.

The soil kind should be loamy or clay loam as water retentivity in such soil is good. Excessive decomposing matter in the soil should not be present at the bottom level. Required quality of good productive soil is like this :

Soil pH	:	6.5 – 7.5
Organic Carbon	:	1.0 – 7.5
Nitrogen (N)	:	30-50 mg/100 g
Available Phosphorus	:	6.0-16.1 mg/100g

Likewise, the quality of water in respect to physico-chemical features should preferably be with the following range :

pH	:	7.0-8.2
Dissolved Oxygen	:	4-10
Turbidity	:	Below 20 ppm
Total Alkalinity	:	75-150 ppm

The nursery ponds are shallow usually with 2.5 to 3.6 feet water (Fig. 10.1). As such, they are seasonal ponds and dry up completely during summer. This situation turns boon as the fish predators get eradicated and unwanted weeds dry up. Drying of such water bodies in the summer naturally helps for mineralisation and also removal of excess of reducing gases. If mud in excess is present, mud from the top layer above 10-15 cm should be removed and it can be used for covering the top layer of the bundh and strengthening it. During rain its utility as organic manure enriches the pond and its productivity is remarkably increased.

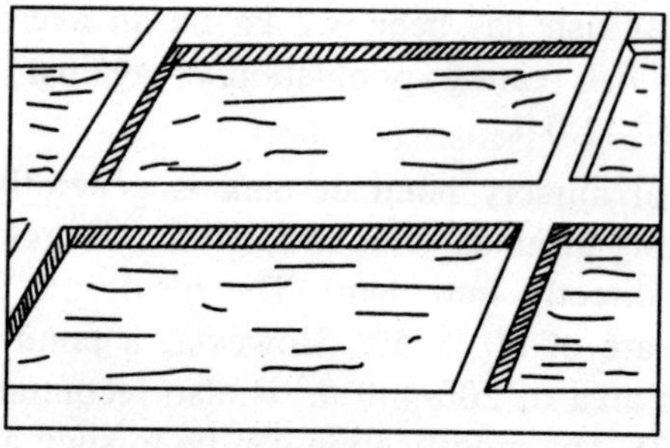

Fig. 10.1. Scenery of Nursery Ponds.

Water from any source can be used for refilling nursery ponds after passing it through a sieve to get rid of predators and insects in any stage. Most common practice is that the nursery ponds are watered 10-15 days before stocking, initially to 60 cm as subsequently raising it up to 100-120 cm.

Eradication of Predatory and Unwanted Organisms

By dewatering and, then, doing water supply, the nursery ponds can be made free from most harmful organisms. However, perennial ponds, *e.g.* small village ponds may be harbouring, besides frogs and snakes, predatory fishes (e.g. *Clarias* spp., *Channa* spp., *Glossogobius* spp., *Mystus* spp., *Wallago* spp., and also weed fishes, e.g. *Puntius* spp., *Ambassis* spp., *Esomus* spp., *Aplocheilus* spp.,) The latter organisms compete for food and prey upon the hatchlings and fry present in nursery ponds.

Control of unwanted organisms, *e.g.*, aquatic weeds, frogs, snakes, predatory fishes and weed fishes is done by draining the pond water through an outlet, or by application of chemicals or selective toxicants including pesticides. The most common measures in this direction are the following :

1. Mahua oil cake. This oil cake is cheap and easily available. It contains 4-6% saponin (a toxicant). Its application at 250-300 ppm kills all the fishes within 6-10 hours. Toxicity persists for two weeks and later the deposits serve as an organic manure in the pond. The special feature is that the treated fish can be consumed by people without any risk of life or ailment (Bhatia, 1970; Nath, 1979).

2. Tea seed cake. This item of tea (*Camellia sinusis*) at a dose of 75-100 ppm kills all the fish in the pond within few hours. The toxicity of this cake lasts for 10-12 days.

3. Bleaching powder. Bleaching powder (Calcium hypochlorite) is also used as a fish poison. It is effective at a concentration of 10 ppm. The toxicity of this chemical generally lasts for a week in the pond. It reacts to oxidise the decomposing organic matters in the pond and indirectly is helpful in controlling fish diseases.

The estimation of quantity of toxicants to be applied in grams is done by multiplying the volume of water in pond (in cubic metres) with the dose (in gram).

Supply of Water

In nursery ponds water from any known source, free from toxicants/contaminants should be filled with after passing it through a fine sieve to get rid off predators and insects in any stage. The water filling work should be done about 10-50 days before stocking. The level of water attained should initially be 60 cm and subsequently it should be raised upto 100-120 cm.

Pond Fertilization

As per finding by fishery scientists including Alikunhi (1955) one of the major factors determining survival and growth of carp fry in nursery pond is the presence of proper kind of natural food in sufficient quantities. It has been generally observed that in nursery pond, where stocking of the fish seed is done at high density, the normal overall food preserve such as plankton (mostly zooplankton) gets consumed in a very short time and the replacement of the food reserve is not possible in a natural way to maintain continuity of the availability. It is urge of the situation to apply manures in such a way that enough plankton growth takes place and the trend continues for a long period. For this aim, the planning of adding manures is done in advance so that there may be high planktonic production nearly synchronising with fry stocking. It is to be acknowledged that after manuring with organic fertilizers, generally zooplanktons in high density appear after first twenty days, secondly that nursery ponds fed from rainwater or other source of supply water do not posses sufficient planktonic population. As such fertilization of the pond is the only course left for the production of sufficient planktonic organisms.

Before taking up manuring work it is required that water and soil conditions should be made stable. In this connection, lime treatment @ 300 kg/ha is done after 2 weeks of eradication of predatory and weed fishes. The action of lime treatment is beneficial in several respects.

1. Acidity of water is corrected.
2. A strong pH buffer system is established.
3. Decomposition of organic matters or deposits on soil takes place.

4. Liming precipitates harmful metallic compounds.
5. Liming serves as disinfectant against fish diseases.
6. Aeration improvement in humus occurs.
7. Resultantly, in the presence of calcium as a nutritive agent, high production of phytoplanktons and zooplankton gets stimulated to fulfil the need of nutrition by growing fish fry.

The methods of use of manures and fertilizers are as per following directives :

1. Application of organic manures. Nursery ponds are manured with raw (fresh) cow dung (R.C.D.) at the rate of 10,000 kg/ha of water surface area about 15-20 days prior to the fixed stocking date. The manure should be thrown all over the pond stocking and specially along the marginal waters.

If it is found that growth of planktonic population is not sufficient after stocking, the pond is further fertilized at a lower rate of 5000 kg per hectare. This second attempt helps in maintenance of the standing planktonic biomass.

2. Application of organic and inorganic manures. The other alternative of manuring is the phased manuring of combination of organic and inorganic manures/fertilizers in two instalments.

1. 5000 kg/ha of raw cow dung and 500 kg of ground nut oil cake or mustard oil cake and 250 kg of ha of single superphosphate are mixed together along with water so as to get a thick paste. The latter is spread uniformly all over the surface of water five days before the fixed date of stocking spawn in the nursery pond.

2. 250 kg/ha R.C.D. 250 kg/ha of groundnut oil cake are mixed together along with water to prepare a paste and the same is applied all over the surface of water two days before the fixed date of stocking of spawn.

A bonus aspect of practising these method of manuring is that the manuring programme can be coincided with the induced breeding practice to get yield of spawn. Secondly, the emerging time of flourishing planktonic population can synchronise with the stocking of spawn in the nursery pond.

Control of Insects

It is commonly found in nursery ponds that aquatic insects and their larval stages live in plenty. They cause destruction of hatchlings or fish spawn in a large scale. As such, the survival of fry in a nursery pond can be assured only after eradication of these harmful insects. Besides incumbent direct loss of life of tiny hatchlings, they may compete for food. Among these organisms, water bugs (*Belostoma*, spp. water scorpions (Nepa), beetles (*Cybister*), back swimmers (*Anisops*) and stick insets (*Ranatral*) are the most common (Fig. 3.6). For the control of these unwanted organisms, several chemicals are known by experimental studies, to be effective (Pakrasi, 1954; Ganguli and Mitra, 1961; Shirgur and Kewalramani, 1967).

1. Use of soap oil emulsion. A simple method adopted for the eradication of insects is by the application of an emulsion prepared with any vegetable oil and cheap soap. The logic behind this relation of chemical is that the aquatic insects respire through the trachea by taking in atmospheric oxygen, which, if obstructed from the water surface, will cost their life by suffocation. Thus the introduction of the emulsion produces on oil film over the water surface which disallows gaseous exchange possibility. The preparation of emulsion is done by mixing 18 kg cheap washing soap and 56 kg vegetable oil respectively. The mixture is heated for a short time and applied 12 to 24 hours before stocking yet uniform film formation over the water surface. A bright aspect of the treatment is that there is no adverse effect on carp hatchling and spawn.

2. Diesel oil emulsion. In recent years, use of low speed diesel oil has replaced the vegetative oil effectively. 50 litres of low speed diesel oil is mixed with 50 ml emulsifier "Hydroxide 1011" and 20 litres water to fulfil the requirement of one hectare water surface area. The emulsion of these produced is spread over the pond as mentioned for the soap oil emulsion. This emulsion is cheaper and shows better results than the soap emulsion.

Maintenance of Water Quality

Wide fluctuations in the physico-chemical parameters of the nursery pond water will not be suitable for the survival of spawn or fry and sometimes, extremely in a dense situation, the mass mortality may take place. Besides, certain physico-chemical factors have untoward effects on planktonic growth so that nutritional proficiency will get affected.

Nurseries with shallow water are relatively free from thermal stratification. Water temperature of warm water region may range between 25° to 30°. On following instruction of culture techniques, the physico-chemical condition of water may be in nearly stable conditon, *i.e.*, are expected to vary within limits. The range of other important physico-chemical condition in nursery ponds should be like this :

Parameters	Range	Parameters	Range
Turbidity	100-180	Total alkalinity	100-250 ppm
pH	7.5-8.5	Phosphate	0.2-0.4 ppm
Dissolved oxygen	5-10 ppm	Nitrate	0.06 to 0.1 ppm
Free CO_2 10 ppm (max.)			

Water in these conditions are considered to be conducive for the upkeep of the nursery pond. It has been observed that normally pond having alkaline water and fair level of phosphate and nitrate favour satisfactory growth and survival of carps fry.

Maintenance of Plankton Population

Growth of plankton in nursery pond is to be evaluated periodically for its quantitative conditions. The method of evaluation is simple. 50 litres of sample of water is collected from different areas of the pond (*e.g.* four corners and one or two middle) and the same is allowed to pass through a plankton net made of butter silk cloth no. 1 (Fig. 10.2). This action stimulates filtration work. By adding 2-3 drops of 0-4% formalin, the plankton is sedimented. Then the sediment is measured volumetrically. A plankton set of 1.0 to 2.0 cc in 50 litres of sample water can be considered as optimum in nursery pond.

The plankton population is generally represented by protozoans, diatoms, rotifiers, cladocerans, etc. These items are considered as items of choice by the developing fish fry.

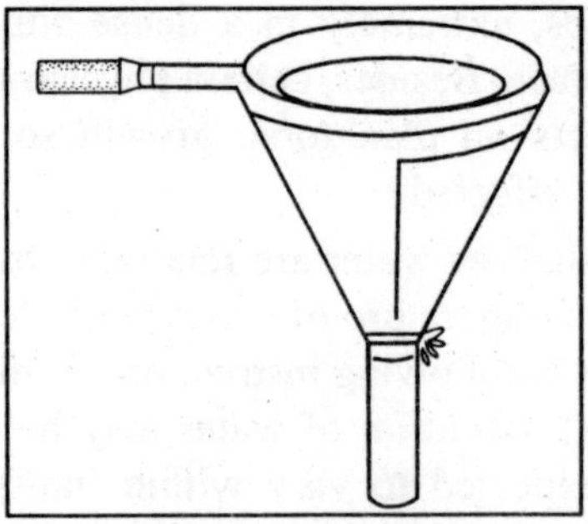

Fig 10.2. A plankton Net.

Stocking

For stocking usual practice is that spawns are transported from breeding or collection centres to nursery ponds. The best stocking time is morning hours in bright sunshine. Late evening time also is an alternative time as the water temperature is relatively low there. The method of introduction of the spawn in the nursery pond is simple. The bucket containing fish spawn is immersed in the pond (Fig. 10.3). By doing so, the spawn gets adopted to the new water media (*e.g.* pond water) and comes out of the bucket in a short time to lead life in new environment.

Fig. 10.3. Stocking scene in a nursery pond.

The experimental results have shown that spawn at the rate of 4-5 million/ha yields good results, *i.e.,* have high survival rate and growth. The farmers of West Bengal introduce 6-10 million/ha in a nursery pond for a period of 5-6 days and later the larvae are transferred to other ponds (Fig. 10.4).

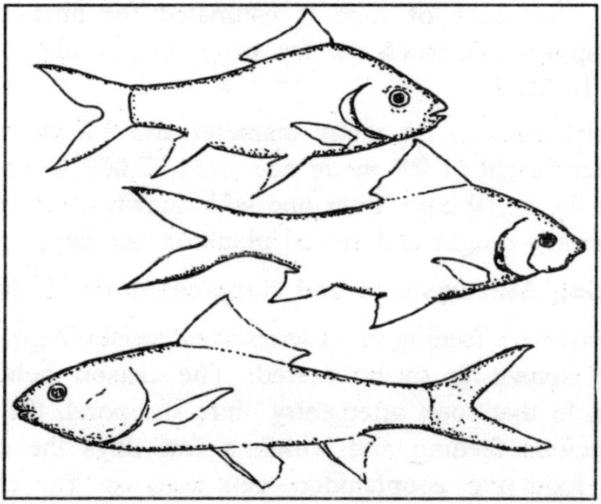

Fig. 10.4. Structure of 3 major carps (adults) most commonly used for raising seeds for pond culture practices.

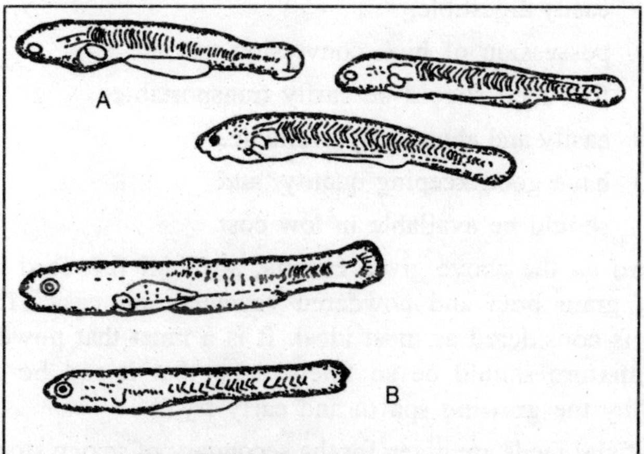

Fig. 10.5. Structure of fry of 3 commonly used major carps : in 48 hours stage (A), 72 hours stage (B).

The density of stocking spawn is an important aspect of nursery management. The amount of food and space available are important points of consideration for deciding the density of spawn. The stock of food is estimated for next 15 days, when the spawn can reach the fry stage, *i.e.*, to about 25 mm size (Fig. 10.5).

A plastic pool of 1.8 metre diameter and 1.2 metre height with a water height of 0.9 metre can yield 77,000 in number of 16-18 mm fry in 10 days from one lakh spawn or fry fed with 5 ml of freshly caught and sieved plankton per day.

Post-stocking Management and Supplementary Feeding

Supplementary feeding in nurseries is essential if high stocking density of spawn is to be reared. The reason behind this supposition is that soon after entry into the pond, the spawns begin voracious feeding and within a few days the stock of planktonic food (*e.g.* zooplankton) gets used up. The choice of supplementary feed is related to these considerations :

(i) readily acceptable to the growing fry,
(ii) easily digestible,
(iii) possession of high conversion value,
(iv) fish food should be easily transportable,
(v) easily and abundantly available,
(vi) have good keeping quality, and
(vii) should be available in low cost.

Based on the above given criteria, artificial fish feed in the form of grain bran and powdered vegetable oil cake (1:1 by weight) is considered as most ideal. It is a must that powdering of the mixture should be so finely done that it can be easily utilized by the growing spawn and early fry.

Artificial feeds are given for the secondary of spawn stocking. The daily ration is estimated on the basis of fry population and average size attained by an individual spawn. An average weight of carp spawn is 1.40 mg and weight of 1 lakh spawn is 140 gram. The total quantity of feed to be fed according to some fishery experts is given below :

Rate of artificial feed	Recommendation by Alikunhi (1975)	Recommendation by Hora and Pillay (1962)
1st 5 days of stocking	Double the weight	Equal to the weight
2nd 5 days of stocking	Three times weight	Double the weight
3rd 5 days of stocking	Four times weight	Three times the weight

To increase the nutritive value of the fish or to have better growth and survivality chance, the fish is fed with a mixture of vitamins, amino acids and trace elements at various compositions and combinations. A close observation has shown that all the feeds spread on the nursery pond water surface is not utilized by the tender early spawn or fry. Rather, some quantity of feed gets immersed in water and settles at the bottom of the pond. After some time they get decomposed and form valuable organic manure. It has been marked that among Indian carps mrigal utilizes the artificial feed in highest extent, but catla does not utilize the artificial feed to any noticeable extent. Furthermore, among the exotic carps, common carp is the best utilizer.

New kind of artificial feed. A recent suggestion has been in favour of preparing a mixture of dry finely powdered notonecta (insect) small prawns and cheap pulses in the ratio of 5:3:2. This powdered mixture is applied at the same rate as that of grain bran and vegetable oil cake. The result of this artificial feed is encouraging (Lakshmanan *et al.*, 1966, 1967; Chakraborty *et al.*, 1979). The growth attained by the fry is generally between 20 to 30 mm and the survival chance is between 88%, as highest, and 50% as lowest.

THINNING AND HARVESTING

In most cases it has been found that the fry grows to 15-20 mm in size after 10-15 days, provided their requirements were fulfilled as per scientific procedure. The rise of population in respect to size necessiates thinning of population. After 10-15 days of rearing, thinning process is undertaken. By using dragnet, at least 40% of the fry should be harvested for transfer to other ponds. This device leads to providing more space and saves food ration for rest developing early fry. If this is not done, further growth of the fry will be hampered and overcrowding may invite other problems.

With normal follow up of methods of fish culture the fry grows to 20-30 mm size in about two weeks. They are finally harvested with fine meshed (1/16 inch) cotton drag net in the morning hours when water is cool and is calm. In a normal breeding season (from July to September, about three crops of fry can be harvested and transferred to rearing pond.

11

Rearing Pond Management : Fingerling Production

Fish culture is practised in a variety of enclosures and in different ways in different ecosystems, yet the village ponds are most popular in rural areas. They differ in shape, size, topography, water supply, water qualities, nature of soil, kinds of natural food organisms, nutrients condition and overall potentiality to yield fish products. From the experiences of fish culture practices, gained over decades, scientific basis of fish culture has been evolved and for any construction and preparation of new culture ponds, the expertise given by the fishery experts is now followed.

During last few decades, India has made remarkable progress in evolving and popularising a technique of composite culture in which high yielding combinations of carps, comprising three Indian major carps (catla, rohu and mrigal) and three exotic carps (silver carp, grass carp and common carp) are allowed to grow in the same pond. Management of fish culture in fresh water ponds needs attention in several respects, such as the preparation of pond, a balanced application of organic and inorganic fertilizers, eradication of weeds and other aquatic vegetation and predators like insects and some fish, stocking the fast growing fish seeds in properly estimated quantity and in appropriate rates and ratio, provision of supplementary feed, production of fish seeds through induced breeding technique, the management of the nursery, rearing and stocking ponds for rearing fry to fingerling and finally adult form (Fig. 11.1). The management of the rearing pond for the production of fingerlings involves much care in the application of techniques so that a satisfactory level of output from the rearing pond can be transferred to the stock pond for getting a rich harvesting. Presently the strategies to be followed for rearing the fries in the rearing pond are discussed in detail ahead.

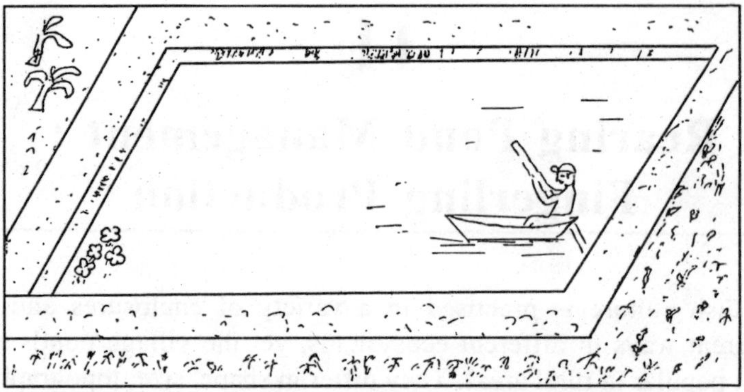

Fig. 11.1. A scenery of a rearing pond.

Rearing of fry to the fingerling stage in a rearing pond, before transferring them to the stocking pond, is considered as essential so as to provide suitable conditions in the rearing pond, *i.e.*, to provide good opportunity of normal growth, fear-free movement in water and for full protection from enemies. As the food and feeding habits of the young fishes (fingerlings) are generally different from those of adults, special arrangement in the rearing pond is urgently needed. It has been observed that the fresh water bodies stocked with fishes of various kinds, such as in the reservoirs the fish yield in kg/ha is never as high as from ponds. Based on observations, specific cultural techniques in rearing the young fry to the fingerling stage have been evolved by fishery scientists in various countries, suiting to the local conditions.

Management of the Rearing Pond

The overall size of the rearing pond is larger than the nursery ponds but smaller than stocking pond. The usual size of the rearing pond is 0.08 to 0.1 ha in the water surface area, and optimum depth of water is between 1.2 to 1.5 m (Fig. 8.3). The method of stocking the rearing pond is much similar to that of nursery pond (Fig. 10.3). The basic technique of the pond preparation, *viz.* pond upkeep or maintenance of suitable optimum quality of water fertilization or manuring are very much similar. The time of rearing of fries is 2 to 3 months; in rare cases upto 6 months. After about 3 months the fry grows to the fingerling stage. Some fishery experts including Lakshmanan *et al.* (1971); Sinha (1973) have suggested manuring course in the following order :

(1) 8 days prior to stocking of fries, organic manuring with raw (fresh) cow dung is done.

(2) After a month of initital manuring a secondary manuring is done with inorganic manures, namely ammonium sulphate, single superphosphate and calcium ammonium sulphate, single superphosphate and calcium ammonium nitrate at the ratio of 11:5:1. The rate of inorganic manuring (*i.e.*, fertilizer treatment) is 1000-1500 kg/ha/yr.

(3) For three months of rearing, secondary inorganic manuring is done at the rate of 500-750 kg/ha. This is preferably done in small dosages, *i.e.*, in a number of instalments so that better results can be secured than manuring at a time.

It has been reported that only inorganic manuring with fertilizers is done in other countries, whereas in India, the condition is slightly different. The organic manures are available in plenty in very cheap rate. As such both organic and inorganic manuring of the rearing pond is done.

As the fry grows to a reasonable size, soon after socking in the rearing pond, a capacity of self-defence develops in them and no control measures for eradication of predaceous insects are considered necessary.

Rate of stocking. The rate of stocking varies widely, depending on the estimation of requirement as per condition of the size, water conditions, food availability and some else factors (Sen, 1976; Jhingran *et al.*, 1979). For a pond with optimum conditions, the density of fry ranges from 2 to 3 lakh/ha in the following combinations and ratio of component fishes :

Species	Ratio
1. Catla + Rohu + Mrigal	3 : 5 : 2
2. Silver Carp + Grass Carp	1 : 1
3. Catla + Rohu + Grass Carp + Mrigal	4 : 3 : 1.5 : 1.5
4. Silver Carp + Grass Carp + Common Carp + Rohu	3 : 1.5 : 2.5 : 3
5. Catla + Rohu + Mrigal + Common Carp	3 : 4 : 1 : 2

The average percentage of survival of the four component carps in the above given 5th alternative has been observed by some scientists to be like this :

Rate of stocking per hectare	Catla	Rohu	Mrigal	Common Carp
62,500	87.51	87.66	91.66	68.40
93,750	84.18	87.47	85.20	63.13
1,25,000	71.35	83.65	77.82	66.98

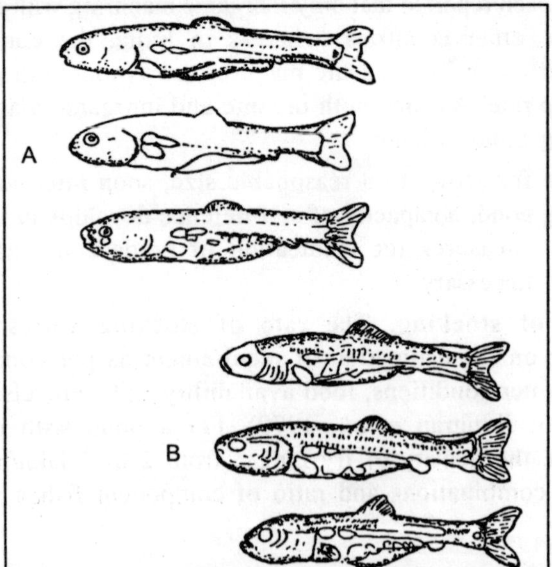

Fig. 11.2. Outline view of fingerlings of 3 major carps : Those in 6th day after hatching (A), those in 7th day after hatching (B).

The fry stocked in rearing pond grows to fingerling size generally after 3 months (Fig. 11.2). This is possible generally as a result to proper feeding. The supplementary feeding is done in the form of mixture of grain bran (rice or wheat bran) and oil cake (mustard oil cake or groundnut oil cake, at 1 : 1 ratio by weight in powder form. The usual practice is to give this powder in the pond everyday during morning hours. The quantum to be given is the following :

(1) One suggestion is like this :

Month	Amount of Feed Mixture to be given
1st month	Equivalent to the intitial weight of the fry
2nd month	Double the equivalent weight of the fry stocked
3rd month	The same

(2) Another suggestion is like this :

Month	Quantity of feed mixture each day/lakh fry
1st month	6 kg
2nd month	10 kg
3rd month	15 kg

The artificial feeding practice may be discontinued if it is found that the water has turned thick green. The development of this colour indicates acute growth of the algal bloom, which is harmful to fish fry, sometimes so much so that the survivality is threatened. One of the noticeable aspect is that they themselves consume dissolved oxygen contained in water for respiration and reduce the availability of oxygen to fish. In the case of very heavy algal bloom the phytoplanktonic activity consume more oxygen than they produce. This situation on cloudy days in particular becomes most gloomy due to obstruction in photosynthesis.

Harvesting of Fingerling

Attainment of the growth of carp fry to a desirable size is dependent on many factors. Therefore, the size actually attained cannot be known without sample checking. It is done by some hand-operated lift net at least once in a month. After completion of 3 months, if it is found that the fry has grown to fingerling size, *i.e.*, attained 100-150 mm size, the same should be harvested repeatedly by drag netting in the morning hours, when water is cool and calm. The harvested fingerlings within no time are transferred to some large stocking ponds where they will grow further to table size.

12

Stock Pond Management : Fish Production

Pisciculture in village ponds and tanks, though on a limited scale, has been practised in India since as early as 350 B.C. This practice of fish culture is said to have originated in Asia sometime in the distant past when fish were raised mainly for ceremonial religious purposes. Nowadays fresh water fish farming has been one of the major occupations of a sizable section of the people of India, particularly of West Bengal, Bihar, Orissa and Andhra Pradesh. Everyday hundreds of trucks loaded with fishes, particularly major carps, *e.g.* catla, rohu, mrigal, common carp, are despatched to the main marketing centres like Kolkata (Howrah), Siliguri, Delhi, etc. In other States too, there has been a marked awakening of interest in the inland fish culture practices. Related to this, there has been a common realization that the fish plays a very important role in increasing the much needed supply of protein to the rapidly growing populations because the land sources for the production of animal protein are fast shrinking as a result of expanding human needs and consequent related activities.

Pressure on land due to population influx and increase in demand for food in recent years have led to the growth of aquaculture industry and it is gaining importance day by day. Data on fish culture practices, fish yields, earning of revenues after marketing give indication that the inland fish culture has a significant role to play in strengthening rural economy, providing employment and in contributing to national income. It has the potential to add several hundred crores of rupees to the national income and provide employment to a large number of fish farmers and fishermen, their dependents, relations and other nominees.

The present level of development as per available estimates is that out of 36.43 lakh hectares of ponds, tanks, beels, oxbow and derelict water etc., 28.55 lakh hectares have so far been brought under cultivation and the recorded inland production of fish in 1995-96 has been 22.4232 lakh tonnes only. As compared to the fish yields in many South-east Asian countries, the present figure is not satisfactory.

CULTURE OF CARPS IN STOCK PONDS

Fish culture is carried out in a variety of water bodies, but the village ponds are the most popular sites. They differ widely in dimension, shape, size, topography, water depth, water supply, drainage provisions, qualities of water and soil (bottom layer). Smaller and larger ponds, after some structural modifications, are used as nursery, rearing and stocking ponds. However, well planned and designed sets of ponds are as project works with scientific approach to intense fish culture. Desirable suggestion usually followed by the fish farmers is that each pond is provided with an inlet and an outlet usually fitted with the concrete pipes of galvanized iron and with the provision of fine meshed screen to prevent the entry of extraneous animals and escape of fish during rainy season (Fig. 8.1).

A stocking pond constructed purposely is rectangular in shape with gradual sloping banks and with an even bottom (Fig. 8.4). The embankment should be firm, provided with inlet and outlet. In the plain areas of India smaller ponds in the size range of 0.2 to 0.5 ha prove to be ideal for getting good return, as the bigger ponds, in 2 ha or more size, are difficult to manage (Fig. 12.1). The depth of water in the pond should preferably be between 2-3 metres but in no part of the year, it should be less than 1.5 metres.

Preparation of the Stocking Pond

Before stocking a pond with fish seeds, some basic requirements are to be met with. The excess of aquatic weeds must be removed. However, some of the submerged rooted vegetation may be left in the pond so as to serve as substratum or additional surface for the attachment of the fish food organisms. For this purpose the common desire is to spare some submerged plants like Hydrilla, Vallisneria, Chara, Najas etc.

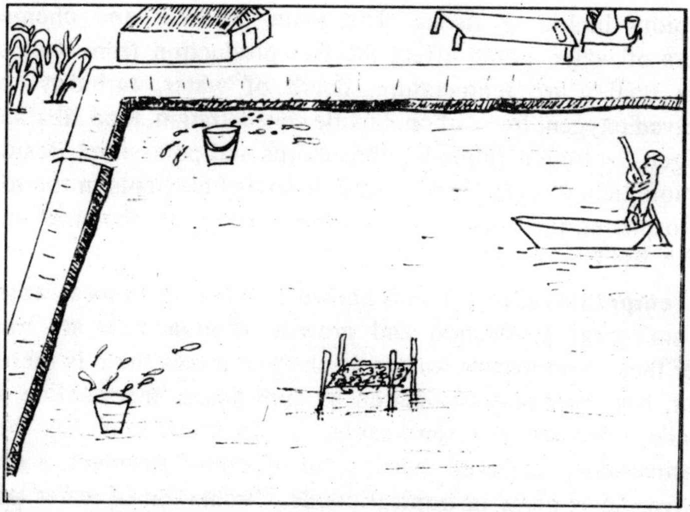

Fig. 12.1. A scenery of stock pond.

The predaceous fishes like *Wallago attu* (a catfish), *Channa marulius* (a giant murrel) should be eradicated. Other undersirable weed fishes like carp minnows and others should also be removed by the use of suitable chemicals as discussed in details ahead in the chapter 'Control of Aquatic Weeds'. In this respect, benzene, nitrobenzene, xylene and turpentine are selective poisons to eradicate unwanted air-breathing fishes and aquatic insects.

Among other requirements of the stock ponds are some physico-chemical factors such as water quality temperature, depth of water, turbidity, hydrogen-ion concentration, dissolved biological oxygen demand, free carbon dioxide, total alkalinity, dissolved nutrients etc. Significance of these and some other factors have already been discussed before in the chapter "Quality of Pond Water and Soil". However some additional informations are given here.

Water Quality

A proper quality of water is prime necessity for fish culture, as its specific characteristics as a culture medium has a direct concern with the productivity of pond water and overall health of the biota including fishes. The main physical and chemical factors of water which affect the fish production from enclosed water bodies are temperature, depth of water, turbidity, pH, dissolved oxygen, free carbon dioxide concentration, total alkalinity, nitrogen, hydrogen sulphide, phosphorus and potassium. Besides, the biological oxygen demand and dissovled nutrients in the form of phosphates and nitrates are important factors for the production of fish food in water.

Temperature. It is a well-known fact that in tropical waters, the biological production and growth of organisms are much faster than in temperate waters. In shallow ponds there is thermal effect, but thermal stratification as epilimnion, thermocline and hypolimnion are not detectable. In Indonesia, it has been experimentally observed that instead of annual turnover, a daily turnover takes place in tropical ponds. Circulation of water takes place during nights with the result that there is a thorough mixing of water and dissolved oxygen and nutrients get well circulated. With the increase in temperature, all life processes in water get enchanced. Biological production and growth of organisms are much more in tropical waters. The oxygen content of the water (dissolved oxygen) decreases proportionally with a rise in temperature. Indian major carps usually tolerate wide range of temperature, but during winter, the growth rate is much retarded.

Depth of water. In respect to nursery ponds (in which hatchlings are reared to fry stage) and rearing ponds (in which the fries are reared to fingerling stage) the ideal size of the pond and water depth should be like this. A pond of 1 hectare size meant for rearing table size fish should have a nursery pond of 0.02-0.06 ha size with a water depth of 1.5-2 m. It has been observed that low deep water reduce normal production due to accumulation of poisonous gases, as the sun light do not reach the bottom of soil.

Turbidity. It is an important limiting factor in the productivity of a pond. If the water is clear, *i.e.*, not turbid, light can penetrate deeper into clear water and influence the photosynthetic activity of plants, migration of food organisms and also induce the growth of plants in deeper waters. Reverse to this condition in turbid water, such as observed in deep water due to the abundance of plankton and suspended fine particles, silt, the photosynthetic activity of plants will be reduced and the fish fauna needing clear oxygen-rich water will be replaced by a predominantly labyrinth form fauna. Too turbidity will reduce the biological production.

Hydrogen-ion-concentration. The pH condition is generally considered as an index of the water condition in a pond. The hydrogen-ion-concentration of a water media below 7.0 value (*i.e.*, acidic condition) is much less productive than alkaline water. Acidic condition of water directly affects the fish life and it may affect the pond life as a whole by impeding the circulation of nutrients, as the rate of decomposition of the organic deposits gets retarded and there is inhibition of the nitrogen fixation. A slight alkaline reaction greatly helps in the conversion of organic matters into assimilable substances, such as ammonia and nitrates. A pH in the range of 7 to 8 is most ideal, being characteristic of good water for fish life as well as other living biota.

Dissolved oxygen. Great significance of dissolved oxygen lies in the fact that fishes and their food organisms get their oxygen for breathing from the air dissolved in water. Aerobic bacteria are also major consumers of dissolved oxygen. During the process of photosynthetic activity of the water plants, oxygen is released and the water gets enriched in the concentration of dissolved oxygen. In tropical ponds, the rapid water circulation enhances the process of aeration. In general, a concentration of 7 to 8 ppm at a temperature of 20°C is considered as ideal for the life of fish and other aquatic biota. Most major carps have been found to sustain even a much lower concentration of dissolved oxygen. The oxygen level is bound to decrease when water is highly contaminated with organic matters.

Biological oxygen demand. Biological oxygen demand (BOD), brought about by an aerobic bacteria, is an important indication of the amount of organic matter in a pond. The factor is measurable with the help of an equipment, called BOD metre, when organic deposits in large quantity get accumulated at the bottom of pond, BOD may rise; there may be depletion of oxygen and, consequently, mortality of fishes and other air-breathing organisms may occur. Such a situation may arise when the water gets polluted due to access of pollutants. In a normal condition, the pond water may have BOD of the order of 2 ppm only.

Free carbon dioxide. Free carbon dioxide is usually present in varying quantities as a result to decomposition of organic matters, such as rotting of these on account of bacterial actions. Rotting of plants often yields free carbon dioxide. Also respiration of aquatic plants and animals produce free carbon dioxide. A high content of free carbon dioxide in water prevents oxygenation of water and it may act as lethal to fish life. Experimental studies have shown that 15-20 ppm of free carbon dioxide has no adverse effect on the fish life, but a high concentration reduces fish productivity. 5 ml of free CO_2 per litre of water is the lethal limit for fish life.

Total alkalinity. The alkalinity of natural water is due to its content of carbonates, bicarbonates and hydroxides of the alkaline earth metals — Na, k, Ca, and Mg and is generally expressed as calcium carbonate. In most waters, Ca and Mg account for the major part of the alkalinity shown. In cultivable water only bicarbonates and carbonates are present. Both hydroxides and bicarbonates cannot exist/occur together because of the undergiven reaction :

$$Ca\ (OH)_2 + Ca\ (HCO_3)_2 = 2\ CaCO_3 + 2\ H_2O.$$

In a normal water, the presence of free CO_2 accounts for increase in the level of total alkalinity, expressed as $CaCO_3$. When the free CO_2 increases, more $H_2\ CO_3$ is produced. Reverse to this, when the CO_2 level decreases, *e.g.* due to photosynthetic activities, more $CaCO_3$ is formed — an expression of total alkalinity.

Dissolved nutrients. Dissolved nutrients such as phosphates and nitrates are important nutrients which are of direct concern for the proper growth of fish and their food organisms. In normal condition the nutrients get dissolved in the surrounding water and keep up a concentration in the water media. However, with the continued cultivation of fish in pond waters, the natural supply of nutrients in water will get steadily decreased. In such a condition, the maintenance of water condition in respect to nutrient concentration in optimum level is done by periodic introduction of manures. However, an excess introduction of manures may lead to the appearance of parasites and other germ-causing diseases. If the bottom soil has thick organic deposits, the use of organic manures causes depletion of dissolved oxygen at a high rate and production of gases of toxic nature which is liable to inflict damage in the tissues of fish. In view of these possibilities, it is imperative to have chemical analysis of water periodically.

Pond Soil

Pond soil differs from land soil in several respects including the fact that the pond soil is water-logged and a gas phase is almost non-existent. In a cultivable pond, normally top 2 to 5 cm of soil is rich in nutrients and concerned with productivity efficiency of the pond mainly with the growth of benthic vegetation. The bed of pond (bottom top soil) releases nutrient material with the water and thus enriches its productive capacity. It has been experimentally observed that the colloidal materials from the top muddy layer help in the fixation or to chemically bind the nutrients (Banerjee, 1967; Matida, 1967). In this respect a noticeable fact is that the nutritive quality of the pond usually gets decreased due to repeated use of the pond for fish culture. The remedial device is that there should be practice of draining fish ponds and the production capacity of the pond bottom should be preserved by alternating periods of mud formation and mineralization with drainage and fresh water supply.

Pond Fertilization

The natural fertility of pond gets reduced on account of continuous growing of the primary and secondary producers. In such a situation, manuring from external sources becomes necessary. After studying the quality of the pond water and bottom deposits of the pond soil, the schedule of manuring should be chalked out.

For ponds with highly clayey top soil, both organic and inorganic fertilizers are suitable. If the pond bottom is neither too clayey nor too sandy and contains medium organic content, a combination of both organic and inorganic fertilizers may be used. If the pond has acidic soil with low pH, a pretreatment of the pond with quick lime @ 250-600 kgs/ha will prove useful.

The procedure adapted for adding fertilizers in the stocking pond is not the same. In ponds having low pH, *i.e.* having acidic bottom soil, the pretreatment of the pond with quick lime 250-260 kgs/ha will yield good result. 2-3 weeks before stocking of fingerlings, introduction of proper fertilizer should be done.

After stocking the pond, manuring endeavour in several instalments should be taken up. Organic fertilizers 20-25 tonnes ha/yr. in 7-11 instalments, and inorganic fertilizers of ammonium sulphate, single superphosphate and calcium ammonium nitrate in the ratio of 11:5:1, and @ 1500-1750 kg/ha/yr. in 4-10 instalments should be given. Regarding introduction of organic manure (given after introduction of fish *i.e.*, fingerling in the stock pond), it would be better to dump the fertilizer under water in corners of the pond, or the same may be put in baskets and brought to corners of the pond under the water (Fig. 12.1).

The information about available food reserve in the pond is obtained by assessment of planktons in both qualitative and quantitative senses. On the other hand, analysis of the soil and pond water will give indication of nutrient level and overall physico-chemical condition of the water (Banerjee, 1967; Worbel, 1967; Matida, 1967; and Ghosh *et al.*, 1979).

Stocking of Pond

For undertaking stocking of pond for fish culture, six species of fish, *viz.* catla (*Catla catla*), rohu (*Labeo rohita*), mrigal (*Cirrhinus mrigala*) along with three exotic species of carp *viz.* two chinese carps, grass carp (*Ctenopharyngodon idella*) and silver carp (*Hypophthalmichthys molitrix*) and Bangkok's strain of common carp (*Cyprinus carpio*) are cultured together (Fig. 12.2). The criteria in choosing this combination are mainly two. (i) They all have rapid growth, reaching table size within less than a year of culture period. (ii) They all are herbivorous with some having zooplankton feeding habit (*viz.* plankton feeders). They utilize most of the food organisms present in the pond.

Catla and silver carp are both surface feeders but the silver carp alone is able to effectively utilize the phytoplankton which is abundant in the pond water. The rohu is a column feeder, but the grass carp does not show any competition because of the fact that it has preference for aquatic vegetation. On the other hand, mrigal and common carp are bottom feeders.

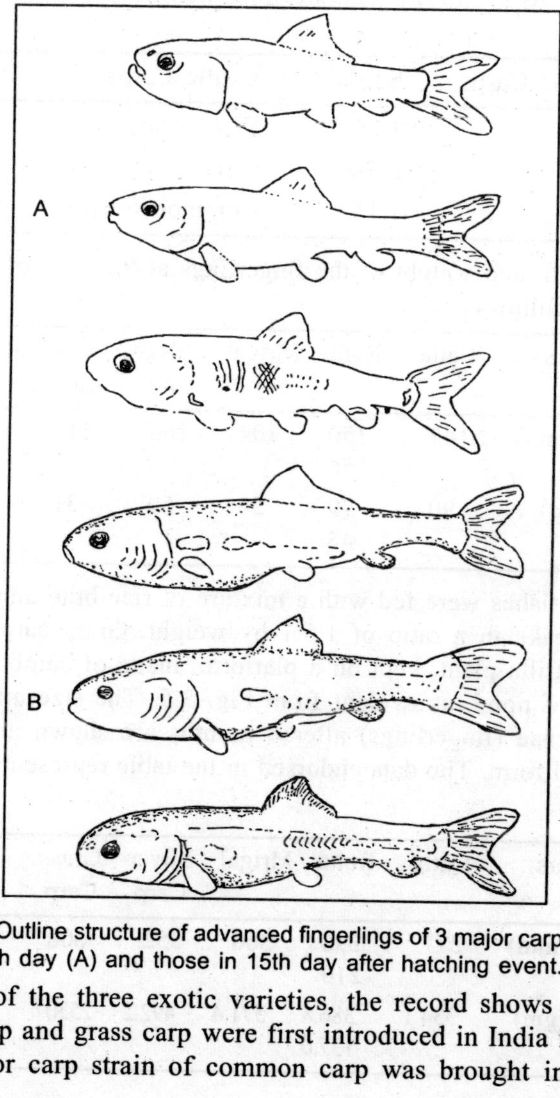

Fig. 12.2. Outline structure of advanced fingerlings of 3 major carps : Those in 10th day (A) and those in 15th day after hatching event.

Out of the three exotic varieties, the record shows that the silver carp and grass carp were first introduced in India in 1959. The minor carp strain of common carp was brought in during

1939. Main features, particularly feeding habits, food choice, breeding time, attainment of maturity have already been given before in chapter 'Culturable Fishes'.

Sinha and co-workers (1973) studied the growth rate of the carps after stocking the ponds with 5,473 fingerlings/ha in the month of April (1972). The stocking was done in the following ratio :

Indigenous Carps	Seeds	Exotic Carps	Seeds
Catla	1.76	Silver Carp	2.29
Rohu	2.74	Grass Carp	0.91
Mrigal	1.15	Common Carp	1.15

Length and weight of the fingerlings at the time of stocking were as follows :

Parameters	Catla	Rohu	Mrigal	Silver Carp	Grass Carp	Common Carp
Length (mm)	147	150 156	104	160	117	44 80
Weight (g)	90	79 45	20	59	31	1 20

The fishes were fed with a mixture of rice bran and ground-nut oil cake in a ratio of 1.5:1 by weight. Grass carp was fed with hydrilla plant, kept on a platform, made of bamboo, in the middle of pond, as evident from Fig. 3.5. The size attained by these larvae (fingerlings) after 6 months are shown ahead in a tabulated form. The data endorsed in the table represents average figures :

Parameters	Catla	Rohu	Mrigal	Silver Carp	Grass Carp	Common Carp
Length (mm)	277	296 210	366	352	608	295 220
Weight (gm)	334.1	386.8 137.0	571.4	492.2	2530	605.6 248.8

Post-stocking Operation

Fish growth and production mainly depend on the management practices, availability of sufficient oxygen, natural food and proper quality of water. With the growth of the fingerlings, demand of food increases. As the stock of natural food is, in most cases, limited, it does not suffice the purpose of the growing young ones. Due to this situation, supplementary feed is necessitated. A large stock can be maintained in a pond with proper and timely provision of supplementary feeds. Before applying feeds, it is essential to estimate the demand of the stock fish. On the basis of size, age group and population figure, quantity of feeds required can be ascertained.

The most common practice is to give carbohydrate rich cereal waste in association with fairly rich protein content. Thus rice or wheat bran and mustard or groundnut oil cake in a proportion of 1:1 serves the aforesaid purpose. The proper time for broadcasting the feed is early hours in the morning and the amount to be given is at the rate of 2.5-10% body weight of the fish stocked.

To fulfil the needs of grass carp in particular, mixing of chopped pieces of green vegetation with the usual feeds and broadcasting the mixture is the usual practice. The growing young ones of the grass carp may have further urge for feeds. Additional feeds like submerged water weeds, duck weeds and land grasses may be given in addition to mixture of cereal waste and vegetable oil cake in 1:1 ratio. Sinha and Gupta (1975) recorded remarkable growth of the grass carp fingerling with sufficient supply of weeds and still better when it was given on wooden platform than dumping them in water (Fig. 3.5).

When it is found that the fish has consumed ali the feeds supplied the feeding schedule should be increased, taking care that there was no waste left over (Fig 12.3). It has been observed that the remains of feeds (unconsumed waste food) gets decomposed, induces growth of thick algal bloom which is harmful, sometimes causing mortality of fish. With this risk in consideration, there should be netting once in two months. The periodic sample netting will indicate the condition of pond water, and, as per need, suitable action may be taken. One suitable way is to re-schedule fertilization of pond or supply of supplementary feeds.

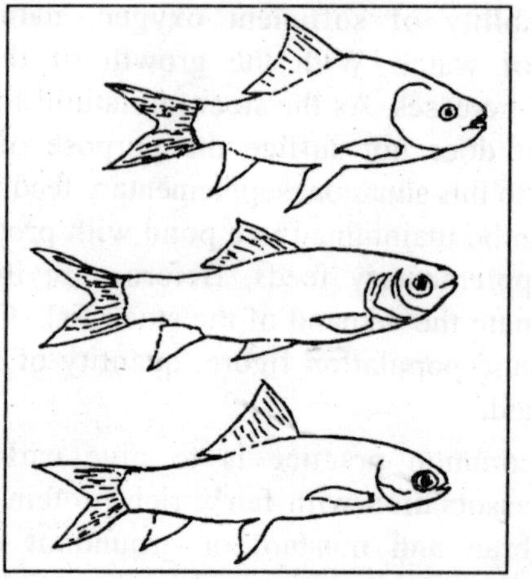

Fig. 12.3. Outline view of 3 young (juvenile) major carps.

REFERENCES

Alikunhi, K.H. and Rao, S.N. 1951. Notes on the early development, growth and maturity of Cirrhina reba (Hamilton). J. Zool. Soc. Ind. 3 (1) : 85-96.

——. 1952. On the food of young carp fry. J. Zool. Soc. India 4 : 77-84.

——. 1956a. Fish culture technique in India. Progr. Fish Developm. India : 63-73.

——, Sukumaran, K.K. and Parmeswaran, S. 1971. Studies on composite fish culture. Production by compatible combinations of Indian and Chinese Carps. J. Ind. Fish Assoc. 1 (1) : 26-37.

Banerjee, S.M. 1967. Water quality and soil condition of fish ponds in some states of India in relation to fish production. Indian J. Fish 14 (1 and 2) : 115-44.

Bhatia, L. 1970. Use of mahua oil cake in fishery management. Indian Fmg. 20 (4) : 39-40.

Chakrabarty, R.D., Sen, P.R., Chatterjee, D.K. and Jena, S. 1975. On the use of fertilizer and supplementary feed for enhancing fish production in freshwater ponds. Proc. Nat. Acad. Sci. (B) 45 (3) : 192-96.

——, ——, and Kowtal, G.V. 1973. Observations on the relative usefulness of different feed for carp spawn and fry. J. Indland Fish Soc. Ind. 5 : 182-88.

——, Rao, N.G.S. and Sen, P.R. 1979. Culture of fish in ponds with fertilization. In : Symposium on Inland Aquaculture, February 12-14, 1979 (Abs.) pp. 5-6. Barrackpore.

Chatterjee, D.K. 1979. Effect of lime on fish pond soils. In : Symposium on Inland Aquaculture, February 12-14, 1979 : 19 (Abs.).

Das, B.C. and Krishnamurthy, H. 1960a. Survival rate of Indian carps, *Catla catla*, *Labeo rohita* and *Cirrhina mrigala* from first to four weeks of life under different experimental treatment. Sankhya B 23 (4) : 301-16.

Devraj, K.V. and Seenappa, D. 1992. Studies on the nutritional requirements and feed formulations for cultivable carps. Inland Fisheries Division, Unit of Agric. Sciences, Hebbal, Bangalore (India). Final Research Report, USDA Funded Project.

Ganguli, D.N. and Mitra, B. 1961. Observations on the fish fry destroying capacity of certain aquatic insects and the suggestion eradication. Indian Agric. 5 (11) : 184-88.

Ganapati, S.V. and Chacko, P.I. 1950. Suggestions for stocking fish ponds in Madras. Madras agric. J., 37 : 1-5.

Ghosh, S.R., Rao, N.G.S., and Mohanty, A.N. 1979. Studies on the relative efficiency of organic manures and inorganic fertilizer in plankton production and its relation with water quality. In : Symposium and Inland Aquaculture (Abstract), February 12-14, 1979 : 107-08, CIFRI, Barrackpore.

Hora, S.L. and Pillay, T.V.R. 1962. Handbook on fish culture in the Indo-Pacific region. FAO Fish. Biol. Tech. Pap. (14) : 204 p.

Jhingran, V.G., Seghal, K.L., Kuldip Kumar and Ghosh, B.B. 1979. Rearing of advanced fry of Indian major carps in recirculatory filtering type of ponds at Barrackpore, West Bengal, Aquaculture.

Lakshmanan, M.A.V., Murty, D.S., Pillai, K.K. and Banerjee, S.C. 1967. On a new artificial feed for carp fry. FAO Fish. Rep. (44) 3 : 373-87.

——, Sen, P.R., Murty, D.S. and Chakraborty, D.P. 1968. Preliminary study on the rearing of carp fingerlings. Indian J. Fish 15 (1 and 2) : 40-52.

——, Sukumaran, K.K., Murty, D.S., Chakraborty, D.P., and Philipose, M.T. 1971. Preliminary observations on intensive fish farming in fresh water ponds by the composite culture of Indian and exotic species. J. Inland. Fish Soc. India 2 : 1-21.

Lin, S.Y. and Chen, T.P. 1967. Increase of production in fresh water fish ponds by the use of inorganic fertilizers. FAO Fish Rep. (44)3: 210-25.

——. 1950. Fish culture in ponds cited by Hora, S.L. and T.V.R. Pillay, 1962. FAO Fish Biol. Tech. Pap. (14) : 204 p.

Motida, Y. 1967. The role of soil in fish pond productivity in Asia and the Far East. FAO Fish. Rep. (44) 3 : 54-63.

Moitra, S.K. and Bhattacharya, B.K. 1965. Some hydrological factors affecting plankton production in a fish pond in Kalyani, W. Bengal, India Khthyologka 4 (1-2) : 8-12.

Murty, D.S., Dey, R.K. and Reddy, P.V.G.K. 1978. Experiments on rearing exotic carp fingerlings in composite fish culture in India. Aquaculture 13 (4) : 331-37.

Pakrasi, B.B. 1954. Preliminary observations on the control of aquatic insects in nursery ponds. Proc. Ind. Acad. Sci. 38B (5) : 211-13.

Ray, P. and David, A. 1969. Poultry manure as potential plankton producer in fish nurseries. Labdev. J. Sci. Tech. 7 B (3) : 229-31.

Saha, G.N. 1969. Relative efficiency of different nitrogenous fertilizers in fish pond soils. J. Indian Soc. Soil Sci. 17(1) : 51-57.

——, Raman, K., Chatterjee, D.K. and Ghosh, S.R. 1975. Relative response of three nitrogenous fertilizers in different pond soils in relation to primary productivity, plankton and survival and growth of *Labeo rohita* spawn. J. Inland Fish Soc. India, 7 : 162-72.

Sen, P.R. 1974. Management of nursery ponds. In "Summer Institute on Inland Aquaculture, 1974, June 27-July 31, 1978 : 6p. CIFRI, Barrackpore.

——, Chakraborty, R.D. 1979. Rearing of major carp fry to fingerlings in fresh water ponds. In : Symposium on Inland Aquaculture (Abstract), February 12-14, 1979 : 18 p. CIFRI, Barrackpore.

Shirgur, G.A. and Kewalramani, H.G. 1967. Studies on eradication of predatory aquatic insects from nurseries. Ind. J. Fish 14 : 215-24.

Sinha, V.R.P., Gupta, M.V., Banerjee, M.K. and Dhirendra Kumar, 1973. Composite fish culture at Kalyani, West Bengal. J. Inland Fish Soc. India 5 : 201-07.

Worbel, S. 1967. The role of soils in fish production in ponds. FAO Fish Rep. (44) 3 : 153-63.

13

Pond Fertilization

The main aim of manuring nursery pond is to induce production of zooplankton which represents the natural food of carp spawn. Fertilizers are used to turn the unproductive state of pond as fertile for boosting production of planktons and other food organisms and enhance growth rate of fish. The effect of application of fertilizers is not instantaneous and also not direct. The ingredients of the fertilizers provide the necessary nutrient substances which are in deficient condition in the pond. The addition of these nutrient substances creates conditions suitable for manifestation of primary producers. This increase in the food matters brings faster growth of the fish and resultantly enhances prospect of harvesting higher yield of fishes.

As said before also, the effect of fertilizers is not at brisk level. The introduced fertilizer is absorbed by the bottom mud which later releases the nutrients in minimal quantities into the water and this condition continues for a long period of time. This quality of fertilizer, *i.e.*, the quality of a prolonged action for a gradual release of nutrients into water, is such that it goes for ever increasing demand of fertilizers to boost up fish production. However, the pond water and bottom soil are to be analyzed first and real needs ascertained about the kinds and quantities of fertilizers/manures. There is no limit to the suggestions made by several fishery scientists from time to time and new informations are coming up about the bright prospects of the fertilizers and manures and also about their imminent adversities if any. Based on these informations, the fish farmers and fishery scientists are in a better position to judge the needs of pond water and take judicious actions regarding the use of fertilizers.

Due to raising of fish crop in the pond, there is consumption of considerable quantities of food items. After harvesting, the pond needs attention for replacement of the food items for new fish crop. The use of fertilizers or manures in adequate quantities can suffice the purpose if real needs are diagnosed in advance in respect to prevalent conditions of water and soil.

Also on stocking a large number of fry in a nursery pond there arises a natural scarcity of food. The availability of natural food in pond is generally not sufficient enough to augment their growth. As such, it becomes imperative to add manures in the nursery pond so that there can be manifestation of zooplanktons which are the natural food of the carp fry.

The schedule of manuring should be prepared after studying the quality of water and pond soil. If the soil is highly clayey, that is, very rich in organic matters, inorganic fertilizer is needed. If the soil is neither too clayey nor too sandy, a combination comprising both organic and inorganic fertilizers may be used, but with medium organic manures. However, if the soil is of acidic nature with low pH a pretreatment with quick lime will be needed so that the water would achieve a neutral or slightly alkaline condition.

The nursery pond needs preparation before hand, *i.e.,* before introducing fingerling in the ponds. The basic needs in this respect are the following :

(1) The pH condition of the pond (water and soil) should be tested. If the condition indicates acidity, it should be treated with adequate quantity of lime so that the condition turns slightly alkaline or neutral.

(2) As all emergent weeds are hindrances for proper growth of fish, give shelter to fish enemies, and consume organic manures so that growth of plankton gets affected, they should be removed.

(3) As floating of weeds interfere with the penetration of sunlight on the water to reach the bottom soil, and affect the growth of plankton, they must be controlled properly.

(4) The pond water should be free from predaceous fishes

such as murrels, catfishes and predatory insects like beetles (*Cybister*), water bags (*Belostoma*), water scorpion (*Nepa*) and backswimmers (*Anisops*).

A removal of these unwanted organisms, by the application of netting and use of proper chemicals as mentioned in detail ahead in the chapter 'Control of Unwanted Organisms', is needed so that introduced fingerling or young ones do not fall prey to these enemies. It is a recognized fact that the nature and quality of fertilizers added affect the species composition of the primary producers and the rate of primary production.

KINDS OF POND FERTILIZERS

Fertilizers applied in fish ponds are of two kinds :

(i) inorganic and (ii) organic. These fertilizers have specialised features for fertilizing ponds and inducing growth of fish food organisms.

Inorganic Fertilizers

The main ingredients of inorganic fertilizers are nitrogen, phosphorus, potassium and calcium. In some other fertilizers, manganese, sulphur, boron, iron, copper and zinc as ingredient elements are present. In this preparation of fertilizers, the required proportion of ingredients is added so that the effectiveness can be as per needs. Most of inorganic fertilizers are soluble in water. Therefore, the elements composing the particular fertilizers are made available to the pond water soon after their application. As most of inorganic fertilizers contain nitrogen (N), phosphoric acid (P_2O_5), potash (K_2O), the expression of quality of fertilizers is generally in the form of percentage composition of these ingredients. Generally the following kinds of inorganic fertilizers are used in fish pond :

(i) phosphate fertilizers
(ii) nitrogenous fertilizers
(iii) lime stone and lime containing fertilizers.
(iv) potassium fertilizers
(v) magnesium fertilizers
(vi) trace elements fertilizers

Phosphate Fertilizers

Phosphate fertilizer is most commonly used for raising fish production in ponds. The use of superphosphate with lime has been found to be much effective in raising fish production. The increase was found by Wunder (1949), on average, to be 96%. It was found that superphosphate was of little value without previous liming of water. Lin and Chen (1967) found the calcium superphosphate alone most effective in raising fish production in the pond of Taiwan.

Nitrogenous fertilizer. Nitrogenous fertilizer of common use are Ammonium sulphate, Ammonium nitrate (Norway saltpeter), Sodium nitrate (Chile nitrate), Ammonia Liquer, Ammonium carbonate, Urea and Calcium ammonium nitrate. Researchers have found positive relationship between fish production and organic nitrogen level, but no relationship exists between fish production and ammonium nitrogen. Another opinion is that there is negative relationship between fish production and nitrate nitrogen. Hickling (1962) has concluded that nitrogen in fertilizer is a poorer fish producer than phosphate. Some others have reported that nitrogenous manures are effective only when there is definite nitrogen deficiency in pond.

Limestone and lime containing fertilizers. Liming of pond is an essentiality, if it is not alkaline, *i.e.*, it is acidic. The quality of lime is generally such that it neutralizes soil acidity and creates a buffer system in the pond water to maintain some sort of stability in the diurnal changes of water from acidic to alkaline condition.

Lime precipitates colloidal matter such as clay suspended in water provides inducement for bacterial breakdown of organic matter. Utility of calcium is in respect to its consumption in the plant growth and also in the bone formation in fish. As lime has toxic and caustic actions, it kills bacteria as well as fish parasites in their various life history stages.

Lime in different forms are used in fish ponds. Quicklime which mostly comprises CaO upto 98% brings drastic changes in ponds in respect to pH, alkalinity, growth of organic matters and, resultantly, plankton production. Next in importance is limestone which contains 50% of CaO. Quicklime can bind much greater quantities of CO_2 compared to the capacity of limestone. However, its adverse effect is that it is a powerful disinfectant and may react as fish poison if in heavy dose. Slaked lime [$Ca(OH)_2$] is highly caustic and, therefore, is not recommendable for use in fish pond. Calcium cyanamide (CaN_2), which contains about 60-70% of quicklime and 18-22% of nitrogen, is not only caustic but also liable to cause death of fish on account of liberated cyanamide.

Potassium fertilizers. Potassium sulphate, potassium nitrate, muriate of potash and kainite are the important potash fertilizers used in agriculture. All potash salts used as fertilizers are water soluble and most potassium fertilizers have negligible effect on the soil pH. The commonly used phosphorous and potash fertilizers affect the soil pH only when they contain nitrogen.

Information on the degree of utility of the inorganic fertilizers in fish ponds is not complete as yet. A number of formulation for the preparation of inorganic fertilizers for manuring fish pond has been evolved and consensus of opinion is that the selection of right kind of potassium fertilizers depends on the soil and water conditions of the fish pond and the same may vary from place to place (Lin and Chen, 1967).

Several experimental results in recent decades have shown that application of 18:18:4 N:P:K @ 5 kg/ha gave satisfactory results in respect to growth of seeds in nursery ponds and of larvae in rearing ponds.

In the stocking ponds, the requirement is different for proper growth of young ones. Ponds are manured with increasing dose of fertilizer combination 18:8:4 :: N:P:K @100 to 500 kg/ha. After a gap of 5 weeks, since the first application of fertilizer, the ponds are stocked with fingerlings of catla and rohu in the ratio of 1:1 @ 2,500/ha. After a gap of 102 days, the average fish production will rise to 68 kg/ha/annum for control and 134, 210, 285 and 443 kg/ha/annum after treatment @ 100 to 500 kg/ha of 18:8:4 :: N:P:K respectively.

Magnesium fertilizers. Magnesium fertilizers have limited utility. It may be useful in ponds with peaty muddy bottom.

Trace element fertilizers. Information on the role of trace elements like zinc, manganese, iron, copper, boron, iodine, molybdenum etc. in the fish culture practices is limited as yet.

Experimental results obtained by Sreenivasan *et al.* (1975) on the plankton production after the use of fertilizers containing trace elements was, however, encouraging. Among the elements tried at 0.05 ppm, zinc gave the highest production, followed by chromium, molybdenum, boron, barium, cobalt, selenium and stannous tin. Banerjee and Banerjee (1967) have additionally provided useful information about fertilization of fish ponds with trace elements.

Organic Fertilizers

Organic manures are generally of composite nature, containing almost all essential nutrients required in the metabolic cycle. After application, these fertilizers release carbon dioxide and nutrients to enhance the fertility of water. An appreciable quality of the organic manures is that they cause release of nutrients to the water at slow rate and prolonged for an extended period of time. In fact, the decomposition of the ingredients of the fertilizers into the ecosystem is in a very slow and gradual rate with the result that the fertility of the soil continues for a long period of time and the growing fish young ones derive, whatsoever, benefit from this situation (Saha *et al.,* 1975).

However, there is some demerit also with the use of organic fertilizers. As they are of composite organic nature, they are liable to carry parasites and other germs causing diseases. In case the bottom soil have thick organic sediment, the use of organic manures causes depletion of dissolved oxygen at a high rate and production of gases of toxic nature which are liable to cause damage to the tissues of fishes.

Following kinds of organic manures have been found useful for enhancing fish production :

(a) Organic manures, containing little or no carbohydrates. Examples : Liquid manures like guano and blood, slaughter house offal and superphosphate.

(b) Organic manures containing some carbohydrate with nitrogen matters. Examples : Farm Yard Manure.

Farm Yard Manures

By the term farm manures is meant the refuse from the farm animals like cow, buffalo, camel, pig, goat, sheep etc. The farm manures comprise the solid as well as liquid forms usually in the proportion of 3:1. On average, more than one-half of the nitrogen, about 2/5 potash and almost all phosphoric acids are found in the solid farm yard manure of cow and pig dung. Dungs from cow-sheds, camel and horse stables contain large quantity of cellulose which may resist decomposition by blanketing the pond soil. Another sort of constrain is with the use of poultry manure. Its use may cause gill-rot in fish. Also farm manures when applied in large quantities may cause deoxygenation and, thus, defeat the very purpose of manuring.

Results of several sets of experiments with the cow dung, superphosphate, sodium nitrate, potassium nitrate in equivalent nutrient bases and in varying doses to study the relation efficiency of manures in nursery ponds showed that only organic manuring gave the best result in respect to yield per hectare.

Use of different combinations of fishes and various fertilizers, both organic and inorganic, showed that the highest fish yield was 4,209.5 kg/ha/yr when the pond at Cuttack were fertilized with 25,000 kg of cowdung/ha (applied in 11 instalments), ammonium sulphate 1,125 kg/ha + superphosphate 500 kg/ha + calcium ammonium nitrate 100 kg/ha, each in 4 instalments. The supplementary feed comprised mustard oil-cake + rice bran in the ratio of 1:1 by weight @ 2,320 kg/ha/year + aquatic weed @ 13,875 kg/ha/year.

Latest view is that fertilization of the pond with organic and inorganic fertilizers in addition to supplementary feeding provides excellent result. However, utility of the nitrogen content is as yet questionable.

Use of sewage. Sewage is considered to be a rich fertilizer with nitrogen and phosphorus as its main chemical constituent. In several countries, addition of a diluted sewage is commonly practised as a pond fertilizer. Particularly in Kolkata, this practice is prevalent with a great success. The application of sewage in pond water is generally in a proportion of 1:4.

Use of sludge. As a fertilizer of pond the sludge is preferred in many counries as it enhances the production of plankton through release of nutrients into water (Natrajan and Verghese, 1979). Its use eventually increases the growth rate of fish in pond water. Use of digested sludge cake is also useful for faster growth rates of catla and common carp in particular. In the activated sludge process, raw sewage is aerated thoroughly for definite periods making the finely suspended and colloid material gather into floccules which gradually settle at the bottom to form activated sludge.

Organic Manures Containing Mainly Carbohydrates

Common examples of the manures in the above category are mustard oil cake, mahua oil cake, soya meal, cotton seed meal, and tea-seed cake. In western countries the fish ponds are fertilized with cotton seed meal, soyabean meal, poultry waste in combination with superphosphate and also of organic manures (cotton seed meal + superphosphate) and inorganic fertilizers (4:10 :: N:P:K). This combination was found there to give better results over organic manures alone.

In India the use of oil cake (mainly mustard oil cake) singly as an organic manure, or in combination with cow dung for manuring ponds is most commonly in practice. The percentage of nitrogen, phosphate (P_2O_5) and potassium (K_2O) in various oil cakes are given below :

Anatomical value (Chemical constituents) of some oil cakes in India

Sl. No.	Manures	N%	P_2O_5%	K_2O%
1.	Mustard oil cake	4.65	1.65	1.56
2.	Linseed cake	5.56	1.44	1.28
3.	Groundnut cake	7.29	1.53	1.33
4.	Mahua cake	2.51	0.80	1.85
5.	Castor cake	4.37	1.85	1.29
6.	Coconut cake	3.02	1.90	1.77
7.	Rapeseed cake	5.21	1.84	1.19
8.	Karany cake	2.97	0.94	1.27
9.	Neem cake	5.22	1.08	1.48
10.	Sesame cake	6.22	2.09	1.26
11.	Safflower (undecorated)	4.92	1.44	1.23
12.	Safflower (decorated)	7.88	2.20	1.92
13.	Cotton seed cake (undecorated)	3.99	1.89	1.6
14.	Coton seed cake (decorated)	6.41	3.89	2.17

Experimental use of oil cakes in Kolkata substation of the CIFRI for fertilizing nursery pond in the form of organic fertilizer (mustard oil cake) + mixed inorganic fertilizer 6:8:4:: N:P:K yielded fast growth of the young ones.

Green manuring. Green manuring is also useful for enhancement of production of plankton and stimulating growth of fish. Most simple method is to sow a nitrogenous or other crop on dry pond bottom before introducing water into it. It supplies organic matter, acts as a food of soil organisms, induces microbial activity in the soil so as to increase the capacity of plankton production, provides conserving influence on the nutrients of soil as it takes up soluble constituents that might be lost by drainage or erosion and, nevertheless, it provides a suitable substratum in the pond for a rich growth of algal matters, zooplanktons, worms, insect larvae which represent the choiced food for fish. Application of loose green manures is not effective unless the loose manures are mixed with mud and then dropped in soil. This additional device prevents chance of drifting away of the loose green manure items and checks chance of deoxygenation of pond water.

Compost manure. Compost manure is popular in many countries including Germany, Indonesia and Congo for fertilizing fish ponds. In India compost manures out of various combinations of cowdung, mustard oil cake and certain aquatic weeds like bladderwort, hydrilla, Naja, Pistia are prepared and used in fish ponds as well as horticulture lands.

REFERENCES

Alikunhi, K.H. 1957. Fish culture in India. Fm. Bull., Indian Coun. agric. Res., 20 : 144.

Banerjee, S.M. and Banerjee, S.C. 1967. Fertilization of fish ponds with trace element manganese for increased production of plankton. FAO Fish. Rep. 3 (44) : 132-52.

——, and Ghosh, S.R. Studies on the correlation between soil reaction and different forms of bound phosphorus in pond soils. J. Inland Fish Soc. India 2 : 113-20.

Bhatia, H.L. 1970. Use of Mahua oil cake in fishery management. Indian Fing. 20 (4) : 39-40.

Chakrabarty, R.D., Sen, P.R., Chatterjee, D.K. and Jena, S. 1975. On the use of fertilizer and supplementary feed for enhancing fish production in freshwater ponds. Proc. Nat. Acad. Sci. (B) 45 (3) : 192-96.

Chang, S.C. and Chu, W.K. The fate of soluble phosphate applied to soils. J. Soil Sci. 12 : 286.

Hepher, B. 1952. The fertilization of fish ponds. 2. Nitrogen. Bamidegeh 4 (10-12) : 220-23.

Hickling, C.F. 1962. Fish Culture : 295 p. London : Faber and Faber.

Hora, S.L. and Pillay, T.V.R. 1962. Handbook on fish culture in the Indo-Pacific region. FAO Fish. Biol. tech. Pap. (14) : 204 p.

Lin, S.Y. 1970. The effect of phosphorus and nitrogen fertilizers on fish production in freshwater ponds. Chinese, American Joint Commission on rural reconstruction, Tai, Taiwan, Fish Series No. 9 : 54-57.

—— and Chen, T.P. 1967. Increase of production in freshwater fish pond by the use of inorganic fertilizers. FAO Fish Rep. (44) 3 : 210-25.

—— and Marzulli, F. 1945. The relationship between the production of fish and the carbon and nitrogen contents of fertilized fish ponds. Trans. Amer. Fish. Soc. 69 (1939) : 257-62.

Mortimer, C.H. 1941. The exchange of dissolved substances between mud and water in lakes. J. Ecol. 30 : 142-201.

—— and Hickling, C.F. 1954. Fertilizers in fish ponds. Fish. Publ. Lond. (5) : 155 p.

Meehean, O.L. 1935. The dispersal of fertilizing substances in ponds. Trans. Amer. Fish. Soc. 69 : 184-88.

Natrajan, M.V. and Verghese, T.J. 1979. Effect of digested sewage sludge cake on plankton and fish production. In : symposium on Inland Aquaculture (Abstracts), February 12-14, 1979 : 101 p. CIFRI, Barrackpore.

Rabanal, H.R. 1967a. Inorganic fertilizers for pond fish culture. FAO Fish Rep. 44 (3) : 164-79.

Saha, G.N. 1969. Relative efficiency of different nitrogenous fertilizers in fish pond soils. J. Indian Soc. Soil Sci. 17 (1) : 51-57.

——, Chatterjee, D.K., Selvaraj, C. and Mazumdar, N.M. 1978. A record of increased fish production in fresh water ponds by use of fertilizers alone. Sci. Cult. 44 (9) : 422-24.

Sreenivasan, A., Pillai, K.V. and Franklin, T. 1975. Role of trace elements in aquaculture. J. Inland Fish Soc. India 7 : 204-08.

Swingle, H.S. 1947. Experiments on pond fertilization. Bull. Ala. agric. Exp. sta. 264 : 34 p.

——, Gooch, B.C. and Rabanal, H.R. 1964. Phosphate fertilization of ponds. Trans. Stheast Ass. Game Fish Commrs. FAO Fish Rep. (44) 3 : 164-78.

—— and Smith, E.V. 1938. Fertilizers for increasing the natural food for fish ponds. Trans. Am. Fish Soc. 68 : 126-35.

Wunder, W. 1949. Fortschrittliche Karpfenteichwirtschaft (Progressive pond-culture of carp) : 386 P.E. Schweitzerbart'sche Veriagsbuch-handlung, Erwin Nagele, Stuttzart.

14

Synthetic Feeds

Presence of natural food in fish pond in adequate quantity is an essential factor in the growth of carps. When the standing stocks of fish in ponds increase over a certain level, the paucity of natural food protein arises. This directly affects fish growth and for the furtherance of growth, suitable supplementary feeds in sufficient quantity are acquired. Another view in the same sequence is that in the intensive fish culture practice, the objective is to maintain optimum density of fish per unit area of water by adopting techniques relating to polyculture, such as multiple stocking, stock manipulation etc. Under such conditions of fish culture, the natural proteins component of food organisms present in the pond water will not meet the needs of the growing fish biomass, thereby necessiating supplementation with protein rich feeds.

A review of old literature reveals that the supplementary feeding has remained an age old practice in Asia and Far East countries where a variety of feeds of both plant and animal origin were additionally used to supplement the deficit of food in fish ponds. In India, however, the supplementary feeding in the ponds of remotely placed rural areas is still in the old pattern, being restricted to use of traditional feeds like rice bran oil cakes. However, the situation has changed in several developed areas where several kinds of artificial feeds and growth promoters have been developed by some fishery scientists including Lakshmanan *et al.* (1967), Das (1967), Bhanot and Gopalkrishnan (1973), Devraj and Seenappa, 1992 and their use as per stipulated situation has revolutionised the prospect of fish yields from pond waters.

In Israel pelleted protein rich diet fortified with vitamins and minerals are being used. Experiments conducted with *Mugil capita* in Israel indicated a percentage increase of weight as given below :

(i)	15.0	with plankton alone
(ii)	127.00	with pellets alone composed of wheat, fish meat, soya flour and 21% protein, and
(iii)	183.00	with pellets and plankton fed together with meat, soya flour and 21% protein.

From these results we can very well conclude the utility of using synthetic feeds along with natural food. It is a considerable fact that the artificial feeds in the form of pelleted vitamin and protein rich diets and bone meal, blood meal, liver, eggs, silkworm pupae, soyabean, maize and potato that are being used in some developed countries might prove either costlier than the fish flesh, or might not be easily available in the remotely placed rural areas of India where the fish culture practices in traditional way are in vogue. As such, some other alternatives were desirable.

In Japan there was practice of young synthetic feeds to carps in running water and in Russia, application of synthetic feeds in floats and cages was an old practice many decades before. This practice began when it was found that the natural foods played insignificant role and laboratory experiments had confirmed that carps could grow well with various types of supplementary feeds so much so that even in the absence of natural food in fish pond satisfactory growth of carps occurred.

In India, the realisation of using supplementary and synthetic feeds came much afterwards and a beginning was made by adding some starchy substance in the living media to serve as dietary calories. Later, several attempts were made to formulate artificial feed for the carps fry. Pioneer workers for this were Alikunhi (1957), Das (1958) and Lakshmanan *et al.* (1967).

Basic research work to formulate new feeds in consideration to kinds of requirement of fish for the proper growth was not taken up in a wider scale and the synthetic feeds formulated was successful mostly for the carp culture only. In the present decade great studies have been made in this connection as in several universities and institutes, researches on this line progressed in right direction and experimentation in controlled conditions gave excellent results. During recent past two decades many new formulations have been prepared and small pellets produced which, on the basis of performance, got recognition from the national and international authorities (Swamy, 1986; Jafri *et al.*, 1988; Lall, 1981; Devraj and Seenappa, 1991). It is a universal understanding that for formulating synthetic feeds, the informations on basic requirements of fish and on the nutritive merits of the various ingredients of synthetic feed are essential.

Merits of Artificial Feeds

Merits of artificial feeds have been reviewed by several fishery scientists including Alikunhi (1957), Das (1958), Lakshmanan *et al.* (1967) and useful informations are now available on the nutrient, and biotic interactions of the artificial feed ingredients and natural food of fishes. Main ingredients in use in India are discussed below.

Fish meal. This represents ideal protein item as all the essential amino acids required in fish feeds are present here.

Silk worm pupae. This represents an ideal feed for intensive farming of carps in floats and cages in Japan.

It shows high rate of efficiency in performance, if silk worm pupae are fed to fingerlings of rohu, catla and common carps along with a mixture of rice bran and mustard oil cake in 1:1 ratio. Chakrabarty *et al.* (1971) reported the usefulness of silkworm pupae feed for fingerlings of Indian major carps and common carp, and recommended for its use for cultivable major carps in general in the fingerling stage.

Plant proteins. These proteins lack lysine and methionine content. Therefore, during application, supplementation with the animal proteins is needed. Various kinds of oil cakes and grain waste have been found quite useful for carp feeding, if given in the proportion of 1:1 in the composite culture of major carps.

Leaf proteins. They are valuable items for fish nutrition. They are cheap, but their use is yet to come up on full scale in India, as there is need of separation of pigments, flavour and toxin which are in fact, unwanted ingredients.

Algal proteins. Nutritional value of this item has been evaluated by several fishery scientists including Singh and Bhanot (1970). Chemical analysis of *Chlorella* spp. have shown the containment of all amino acids. Feed pellets from *Chlorella* have been prepared and marketed in open. Their uses have given high growth rates and a better condition factor for *Tilapia mossambica* and some other fishes (Poma, 1982).

Single cell proteins (SCP). These proteins are derived from yeast, bacteria, algae, or fungi, which are grown on various kinds of strata, such as hydrocarbons (crude oil, gas oil, natural gas, coal), carbons (molluscs, cellulose, grain, sulfite liquor and organic wastes through experimental devices. Successful results have been derived, when yeast, grown on liquid hydrocarbons, was used in fish culture.

Carbohydrates. Carbohydrates in the form of starch diets are used as a major source of dietary calories in artificial feeds. Rice bran and wheat bran, which are common starchy substances are used as carp feeds since long time past. Nevertheless, small sized potatoes, which are cheap as well can be used in place of grain for feeding carps.

Fats. These are not a favoured item for use in feeding carps. They are better utilized for feeding fish cultivated in warm waters. Cod liver oil when added to the diet of some fishes promotes growth. For warm water fish such as channel catfish, 5% corn oil gives satisfactory result.

Micronutrients. These are chemical products commonly known as growth stimulants or growth essentials. This, in fact, is not a food item, but the presence of this item in small quantity helps as stimulant of growth and improves protein assimilation.

Vitamins. Research works have been in advanced stage about the requirement of vitamins for trout and salmon. However, the extent of utility of vitamin for carp culture is known at a limited scale uptill now. Feeding of 0.8 mg/kg/day of cobalt chloride and addition of 4% fodder yeast to carp feed gives better growth rate. Effects of cobalt chloride (for vitamin B_{12}). Yeast vitamin B_{12} and ruminant stomach extract on the growth and survival of the spawn of carp have been studied by several scientists. Results of experimentation done by Sen (1972) showed that the addition of cobalt chloride in the feed @ 0.01 mg/day/fish to fry of *Labeo rohita* led to rise in survival and growth by 8.33%, and 2.32 mm/35.7 mg respectively.

Antibiotics. It is a well-known fact that fishes, particularly, its young ones, are more prone to diseases when their population is raised by culture practice with the provision of feeding on intense level. The wastes left after feeding harbour bacteria and other germs and these infect young one, of fish. It has been found that treatment with antibiotics, like terramycin, yields encouraging result as the mortality rate of the fish decreases. Thus by feeding 20,000 units of terramycin to carp every 3 days results in the growth increase by 9.5% and a saving of 10.5%, as per reporting by Das (1967). As the chemical is costly, its use is on a limited scale uptill now.

The table 14.1 gives necessary information on the nutritive value of various feeds, as per reporting by Lakshmanan *et al.*, 1966.

TABLE 14.1

Nutritive value of different feeds

Feed	% Protein	% Fat	% Carbohydrate	% Ash
Zooplankton	46.40	6.00	22.60	22.50
Silkworm pupae	59.72	6.51	18.97	15.80
Mustard oil cake	24.60	1.20	41.60	6.50
Rice bran	13.20	1.80	78.50	6.50
Groundnut oil cake	32.41	11.43	39.89	16.27
Wheat bran	11.23	2.48	83.96	4.33
Soyabean	37.63	16.46	40.64	5.27
Prawn waste	17.77	3.34	29.62	49.27

As kinds of nutrients are directly related to the growth of fish, that is weight increase of the fish or, one can say, weight increase in respect to fish flesh build-up, it can be inferred from data of the table that the order of nutritive value is, from higher to lower order, like this; silkworm pupae, zooplankton, soyabean, groundnut oil cake, mustard oil cake, prawn waste, rice bran and wheat bran.

Another valuable information in the same respect is available also from the account given earlier by Banot and Gopalkrishnan (1973).

Table 14.2

Feed	% Protein	% Carbo-hydrate	% Fat	% Fibre	% Ash	% Moisture	Authority
Zooplankton	45.65	10.25	4.8	—	5.20	65.80	Lakshmanan
Notenectids	56.2	24.1	4.2	—	15.5	—	*et al.*, 1967
Prawn	18.8	32.8	8.4	—	40.0	—	—
Fish meal (white)	61.0	1.5	3.5	—	21.0	13.0	Hickling, 1962
Silkworm Pupae (dried)	55.91	24.5	5.58	—	2.98	10.03	Hora and Pillay, 1962
Silkworm Pupae (defatted)	75.36	8.40	1.75	—	5.59	8.90	" "
Phytoplankton	40.50	25.30	3.4	2.3	15.20	85.94	
Rice bran	13.24	42.65	4.30	6.50	13.28	9.15	Hora and Pillay, 1962
	15.80	47.40	16.05	6.63	14.80	11.20	
Wheat bran	13.90	55.60	4.20	10.50	5.30	10.50	" "
Barley	10.29	66.34	2.17	5.81	2.36	13.03	" "
Soyabean cake	37.03	9.43	9.43	4.81	4.84	15.86	" "
Mustard oil cake	30.6	29.2	10.22	—	30.0	—	Lakshmanan *et al.*, 1967
Maize	7.31	79.9	4.5	—	8.5	—	" "
Groundnut cake	44.4	22.9	7.2	9.0	4.6	12.1	Hickling, 1962
Potato	2.1	19.7	0.1	0.9	1.0	76.2	" "
Safflower oil meal	20.1	24.7	0.3	39.2	3.5	12.2	Bondi *et al.*, 1957
Fodder yeast	42.52	13.16	2.3	—	6.10% minerals		Sukhoverov, 1967

Based on utilization reports, as given in table 14.2, one can judge the performance of artificial feeds. Indices of utilization can be derived by equating quantity of food given (in any weight) and weight increase of fish (in wet weight). In other words, indices of utilization are derived from the undergiven equation.

$$\frac{\textit{Weight gain}}{\textit{Weight of food consumed}} = \text{Conversion co-efficient or conversion efficiency}$$

A theoretical calculation can be derived on the basis of two established facts :

(i) the feed given to fish usually contain 40 per cent protein.

(ii) However, fish flesh has about 14 per cent protein. Usually 2 kg of feed is given for every 1 kg of fish flesh. Supposing the factor from the protein in the feed of fish to fish flesh is 1:1, the conversion factor between protein in the feed to protein produced by the fish comes to 1:7. Logical inference thus drawn is that with every 7 kg of protein in the feed only 1 kg of fish protein is produced.

It has universally been realised that the supplementary feeding is the most important necessity in the management of stocking pond. By undertaking this task a large stock of food can be maintained in a fish pond and nice opportunity provided to the growing fish to attain rapid growth.

The most popular device is the use of carbohydrate rich cereals wastes *e.g.* rice or wheat bran in combination with vegetable oil cakes (mustard or groundnut oil cakes which contain fairly rich protein content). These two items are mixed together in the ratio of 1:1. The daily ration quota is 2.5 to 10% of the body weight of the fish stocked. The best time of broadcast is early morning when the rising sun effect is nil or least in the pond water.

For grass carp chopped pieces of green vegetation are mixed with the supplementary feeds as mentioned before and spread in the pond. This device gives encouraging results. To meet the additional requirement of this fish, the vegetation like submerged water weeds, duck weeds and land grasses from external sources can be procured and given at times. However, bran and oil cake in 1:1 ratio remains a patent feed for this fish like other fellow fishes in the pond.

It is a rather difficult task to determine the quantity of supplementary feed to be provided. However, a logical estimation can be made on the basis of few observations. When a fish is able to consume all the feed supplied, an understanding comes for the insufficiency of the feed supplied. As such, the feeding schedule should be increased proportionally, keeping in view that there will not be wastage. This caution is based on the fact that the unconsumed feed chopping as waste in pond water, may develop a thick algal bloom which obstructs the free movement of the fish (carp) in search of zooplanktons, and in hot summer days, this condition may lead to mortality of fish due to depletion of oxygen, as mentioned earlier also.

It is well-known fact that the carbohydrate food, after reaching the gut, is subjected to enzymatic actions and converted to diffusible form, that is glucose. The latter gets absorbed in the gut wall and reaches blood stream to be utilized as energy source for various purposes of life activities. However, the excess of glucose is converted to glycogen by the process of glycogenesis and stored in the liver. At times this reserve food in the liver is converted to protein by the complex physiological processes of Embden-Meyerhof pathway and Kerb's cycle.

Thus carbohydrate food generally taken in larger quantity than proteinous food matters (because of less caloric values) have the same ultimate fate as proteinous food and add to growth of the fish and increase in weight per unit area of the fish.

The idea of feeding protein foods to all cultural species does not sound economic in consideration of the prospects of other alternatives whose nutritional values have already been determined. It is a recognized fact that there is a heavy demand of ingredients of protein foods to prepare feeds for human consumption. However, several alternatives have been sought which keep items of human needs excluded from cheaper items of protein foods for use by fish. In this category are plant proteins, condemned grains, straw, hay, scrap of human food, by-products and wastes from fodder industries, single cell proteins, etc. which are not consumed by human beings for quality reasons. The utilization of such substances to formulate artificial feeds after proper investigation and experimental trials can help a lot in the gradual development of a fish feeds industry.

REFERENCES

Alikunhi, K.H. 1957. Fish culture in India. Fm. Bull., Indian Coun. gric. Res., 20 : 144.

Bhanot, Kuljeet K. and Gopalkrishnan, V. 1973. Priorities in Nutritional Research for formulating artificial feeds for fishes. J. Indian Fish. Soc. Vol. V : 162-70.

Chakrabarty, R.D., Sen P.R., Chatterjee, D.K. and Kowtal, G.V. 1973. Observations on the relative usefulness of different feed for carp spawn and fry. J. Inland Fish Soc. India Vol. V : 182-88.

——, Sen, P.R., Chatterjee, D.K. and Jena, S. 1975. On the use of fertilizer and supplementary feed for enhancing fish production in fresh water ponds. Proc. Nat. Acad. Sci. (B) 45 (3) : 192-96.

——, —— and Kowtal, G.V. 1973. Observations on the relative usefulness of different feed for carp spawn and fry. J. Inland Fish. Soc. India Vol. V : 182-88.

——, Murty, D.S., Sen, P.R. and Chatterjee, D.K. 1973. Preliminary observations on usefulness of silkworm pupa as feed for fingerlings of Indian major carps and common carp. J. Inland Fish. Soc. India 5 : 117-18.

Chatterjee, D.K. 1979. Effect of lime on fish pond soils. In : Symposium on Inland Aquaculture, February 12-14, 1979 : 19 (Abs.) Barrackpore.

Das, I. 1958. Observations on the nutrition of the carp spawn of the Mahanandi river. Proc. Nat. Inst. Sci. India 24B (6) : 273-84.

Das, B.C. 1967. Effects of micro-nutrients on the survival and growth of Indian carp fry. FAO Fish Rep. (44) 3 : 241-56.

Devraj, K.V. and Seenappa, D. 1992. Studies on the nutritional requirements and feed formulations for cultivable carps. Inland Fisheries Division, Univ. of Agric. Sciences, Hebbal, Bangalore (India). Final Research Report, USDA Funded Project.

Lakshmanan, M.A.V., Murty, D.S., Pillai, K.K. and Banerjee, S.C. 1967. On a new artificial feed for carp fry. FAO Fish Rep. (44) 3 : 373-87.

——, Sen, P.R., Murty, D.S. and Chakraborty, D.P. 1968. Preliminary study on the rearing of carp fingerlings. Indian J. Fish 15 (1 and 2) : 40-52.

Sen, P.R. and Chatterjee, D.K. 1979. Increased production of major Indian carp fry by addition of growth-promoting substance. Proceedings of the World Symposium of Finfish nutrition and Fish Feed Technology, Hamburg 20-23 June 1978, Vol. 1, pp. 190-95.

Singh, C.S. and Bhanot, K.K. 1970. Nutritive food value of algal feeds for carp. J. Inland Fish Soc. Ind. 2 : 121-27.

Sinha, V.R.P. 1978b. On the contribution of supplementary feeding in increasing fish production through composite fish culture in India. Symposium on fin-fish Nutrition and Feed Technology. Hamburg, 20-23, June, Vol. 1, 1, Berlin, 1979, pp. 567-72.

15

Carp Culture

Pisciculture has been practised in ponds and tanks in India since as early as 350 BC. It is nowadays very popular in the states of West Bengal, Orissa and Bihar. In other states also there has been a marked awakening of interest in the inland pisciculture practices and recently fish farming has come up as a major agricultural activity in Andhra Pradesh, having a well managed marketing system to transport surplus fish to Bihar, Orissa, West Bengal, Uttar Pradesh, Delhi and Haryana. Kolkata, Patna and Delhi are major receiving centres of fish from Andhra Pradesh. Unfortunately in larger inland water areas pisciculture on scientific lines is not practised and the production of fish has been limited, as compared to total areas of inland water resources, which is 171334 lakh hectares as per 1995 survey report. The figures of total yield for recent six years (1990-96) has been recorded as 1536.25, 1709.33, 1789.05, 1995.50, 2096.73 and 2242.32 thousand tonnes only. These are, obviously, not encouraging figures, when compared to figures of per acre yield of several southeast Asian countries.

The greatest advantage of expanding fish farming programmes lies in the involvement of a large population of rural areas in various aspect related to fish farming and boosting their economic condition. This envisaged scope can solve the unemployment problem of the youth to a considerable extent. Secondly, fish being a rich source of proteinous food, provides a great scope of sufficiency in meeting with the protein requirement of the general population. The liver of cod and shark marine fishes carries medicinal value in respect to the fact that their liver has the storage of valuable vitamins.

A scientific method of the inland pisciculture practices needs consideration of several factors including physico-chemical condition of water, and selection of cultivable species of fishes. Salient features of these requirements are given below.

Physical and Chemical Conditions of Water

Important physical and chemical conditions of water affecting the fish productions are depth of water, turbidity, carbon dioxide, acidity, alkalinity, neutrality, dissolved oxygen, dissolved nutrients, etc.

Depth of water. Too deep water reduces normal fish production due to accumulation of poisonous gases in water. The reason behind this fact is that the sunlight fails to penetrate to the bottom of water and to allow manifestation of phytoplanktons and other living organisms.

Turbidity. It may occur in deep water due to the abundance of plankton or suspended fine particles or silt. Too much turbidity (above 180 ppm) reduces the biological production.

Temperature. In tropical waters, the biological production and growth are much more rapid than in temperate waters. The major Indian carps usually tolerate the wide range of temperature. During winter, their growth is very much retarded. Temperature level has direct relationship with the oxygen retaining capacity of water. It has been observed that in the fresh water at 0°C temp. and 30°C temp. the oxygen saturation points in millilitre per litre are 10.27 and 5.57 respectively.

Light. Since the ponds are shallow, the light penetrates to the very bottom. However, abundance of vegetation tends to curtail light in the deeper waters.

Dissolved oxygen. For breathing, fishes use oxygen dissolved in water. Source of oxygen in water is the atmospheric air. However, photosynthetic activity of water plants also enriches the water with oxygen. The oxygen is bound to be deficient when water is highly contaminated with the organic matters, such as heavy infestation with the aquatic weeds and planktons. As they use more and more oxygen available in water for respiration, scarcity may arise in respect to availability of oxygen to the fish in ponds. If there is heavy algal bloom in the fish pond, they use more oxygen than they produce. Such an adverse condition becomes more critical on cloudy days.

Major carps have been found to tolerate even a much lower concentration of dissolved oxygen and they can survive under stress as well. If the oxygen level reaches a very low level, much below the critical level for fish in general, a case of mass mortality may occur.

Carbon dioxide. The amount of free or uncombined carbon dioxide in water is of ecological importance and is related with the productivity of ponds. The decomposition of organic matter and the respiratory activity of aquatic plants and animals produce carbon dioxide. This gas is one of essential raw materials for photosynthetic activity of green plants. It influences the hydrogen ion concentration in water.

Dissolved nutrients. Phosphates and nitrates are very essential for determining the fertility of pondwater, and consequently, for the production of fish in the ponds. In all natural waters, such as ponds, tanks, the nutrients are found in varying quantities in dissolved condition surrounding bottom soil. If nutrients are not sufficient, *i.e.,* there is deficiency of production of the primary organisms and growth of fish young ones in the water areas affected. Several scientists including Alikunhi (1956, 1967), Singh *et al.* (1972), Chakrabarty and Murti (1972), Sinha *et al.* (1973), Chakrabarty *et al.* (1979) have given valuable information about various species of carp culture.

CULTIVABLE SPECIES OF CARPS

Fast growing species of fish with non-predaceous feeding habits are generally selected for cultivation in ponds. Some exotic species of food fishes have also been introduced into this country. Major carps utilized for culture are catla (*Catla catla*), *rohu* (*Labeo rohita*), mrigal (*Cirrhinus mrigala*), silver carp (*Hypophthalmichthys molitrix*), grass carp (*Ctenopharyngodon idella*) and common carp (*Cyprinus carpio communis*). They have following characteristic features :

Catla catla. This is the fastest growing carp in India. It has a characteristically deep body with a broad head, large upturned mouth and non-fringed lips devoid of barbels (Fig. 2.1). Its dorsal fin is broad and has 14-16 branched rays. Its fry is 12.5 to 25 mm in size, and has relatively large head with reddish gills, broad dorsal fin with a greyish margin and pale body colour.

Catla is a surface and column feeder and upturned mouth and large gill rakers are adapted for feeding numerous organisms floating in water. Submerged aquatic weeds are not of any value as food for Catla. It has a moderate growth rate, reaching a size of 37.5 – 45 cm in the first year and to a maximum of over 4 feet in length in the adult stage, when the weight attained is high, being 3-6 kg after 3 years. Fishes reach maturity generally in the 2nd year. Its size at maturity varies highly from habitat to habitat. The males have lower body weight in relation to body length than of females of comparable age.

Catla breeds in the river during the rainy months of June to August. It also breeds in the bundh type tank in Bengal and Bihar. Early fry is available during June-July in Assam, Bengal and Bihar and during July-August in Orissa, Punjab and Andhra Pradesh, and collected in millions and stocked in the nursery ponds.

Rohu. It is very fast growing indigenous carp. The head is small in size, pointed in front, provided with dull reddish scales on the sides and with pink reddish fins. The body is more linear than in catla and the lower lip is fringed (Fig. 2.2). The rate of growth in general is slightly lower than that of catla. The age of maturity is generally 2 years, but the males may reach maturity even in 1 year. A 2-year mature fish weighs between 1.5 to 3 kg.

Its young ones, 20 to 25 mm long, are distinguished by their fringed lips, dorsal fin rays and a dark spot at the base of the tail. However, the latter disappears in the fingerlings. *Labeo,* which is commonly known as 'Rohu' is generally bottom feeder. Young ones feed on water fleas and animalcules like the catla fish, but the growth is relatively slower than catla, reaching generally 35-40 cm in the first year. Sexual maturity is attained by the end of the second year and the maximum size reaches over 3 feet. Breeding takes place in the river and bundh during monsoon.

Mrigal. This fish, in flavour-based importance, is next to catla and rohu and the same grading is also for cultural purposes. It is easily distinguished by the relatively linear body, small head, blunt snout portion, terminal mouth with thin non-fringed lips, bright silvery body, with reddish fins (Fig. 2.3). The dorsal fin has 12 to 13 branched rays.

It attains sexual maturity in 2 years, but on proper care of feeding in fish ponds, the fish can attain maturity even in 1 year. It has been successfully induced bred in ponds. The brood fish, when maintained properly, is likely to attain 1.5–2.5 kg body weight in about 2 years. It grows at slower rate than catla as well as rohu, attaining a maximum length of 75 cm.

Silver carp. This is an exotic fish introduced in India from Hongkong in 1959. This is the fastest growing carp in the pond water and attains full maturity in 2 years (Fig. 2.4). However under proper care the maturity in both sexes can be attained in 1 year also, and they can be used for induced breeding. The males mature in 1 year. The size of the fish at 2 years of age varies a good deal and weighs between 2.5-5 kg. In China, which is its native country, the silver carp takes more time for attaining maturity, *i.e.*, attains maturity after 2-6 years as per reporting by Tapiador *et al.* (1977). In U.S.S.R. the fish matures within 3-10 years and in Rumania within 6-9 years. Usually the males mature a year before. It is predominantly a phytoplankton feeder and in respect to growth, it is a superior fish to catla.

Grass carp. This fish is another very fast growing exotic carp and grows to the size of silver carp, if proper conditions including feeding care are given. The fish attains maturity in 2 years, but the male may be in an oozing condition in 1 year. The brood fish may attain a size of 2-4 kg in 2 years (Fig. 2.5). However, The males weigh less than the ripe females of the same age.

Experimental consignment of this fish were brought to India in the form of fingerlings for the first time in 1959. They have been successfully induced to breed in India and now extensively used for culture in consideration to their remarkable growth and biological activity in Indian conditions. The grass carp feeds voraciously the weeds like *Hydrilla, Lemna, Azolla,* etc.

Common carp. This fish was introduced in India from Sri Lanka in 1939. It is a cold water fish, but is very hardy and easily adapts itself to warm water conditions of tropics. Originally this fish was stocked in the Nilgiri Hills in Tamil Nadu, but it has spread now to other states also and is now selected as a component of composite culture practices. It grows to about 3 feet size in length. It matures even in 6 months and the mature brood fish attains 1-2 kg body weight within a year (Fig. 2.6). The weight of testis in relation to body weight is around 12%, being the highest figure among the cases of all carps. Three varieties of common carp are available in India : (i) the mirror carp (*Cyprinus carpio* var *specularis*) which has large shining yellowish scales over the body, (ii) the scale carp (*C. carpio* var *communis*) with rather small scales completely covering the body, (iii) the leather carp (*C. carpio* var *nudus*) with scales almost completely absent and giving a leathery appearance.

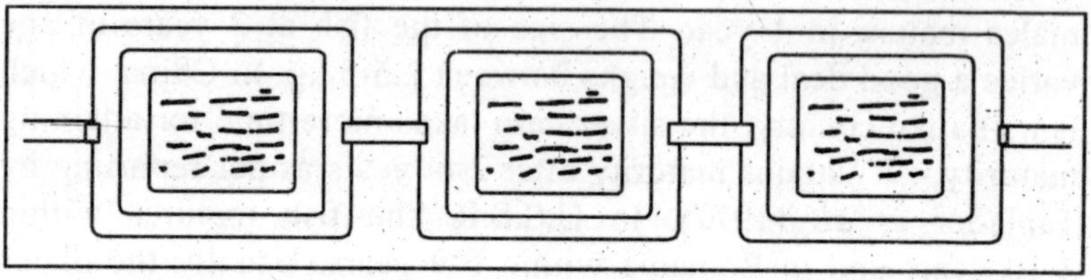

Fig. 15.1. A view of hatching pits, interlinked for regularity in water supply.

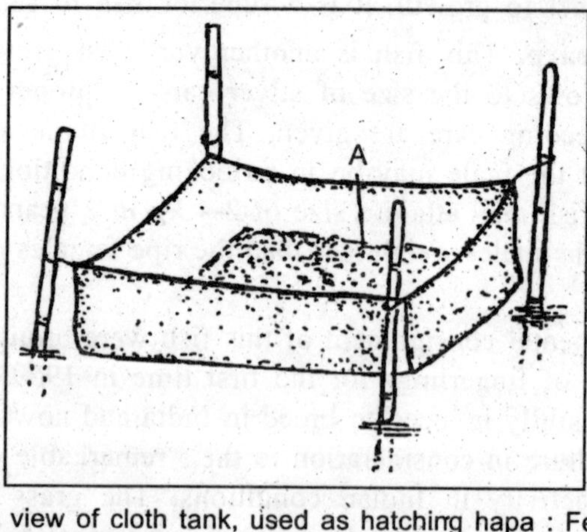

Fig. 15.2. A view of cloth tank, used as hatching hapa : Fertilized eggs marked as A.

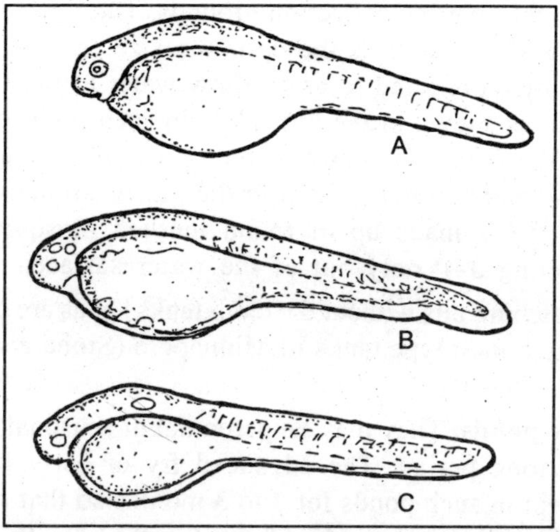

Fig 15.3. A view of hatchlings of 3 major carps : Katla, *Catla catla* (A); Rohu, *Labeo rohita* (B); Mrigal, *Cirrhinus mrigala* (C).

The cyprinus carp is omnivorous in feeding habits and breeds in the pond. The young fry feeds largely on planktonic crustaceans. The eggs are found sticking to the leaves of submerged aquatic plants as evident from the figure 5.6, given before. The young ones hatch out within 58-96 hours in winter and 26-60 hours in March-April. The common variety of the fish chosen for culture practices in the plains of India is Indonesian strain of the common carp, *C. carpio* var *communis*.

CULTURE OF CARPS

The major carps of India do not breed in the confined waters, except in the bundh type of tanks such as observed in Midnapore and Bankura districts of West Bengal (Ganapati and Chacko, 1954; Ghorai *et al.*, 1979). It is, therefore, necessary to obtain their fries or fingerlings for pisciculture from the natural habitats, *i.e.*, rivers. They are first stocked in the nursery ponds for rearing, so that they can gain resistance capacity.

Preparation of ponds. Different methods for the preparation of ponds are followed. The important phases of the carp culture and the types of ponds required for each phase are as follows :

Hatching pits. They are located close to the breeding

bundhs or to the riverine collection grounds. Their size may vary, but generally 8′ × 4′ × 2′ is the optimum size. A series of such pits remain interconnected so as to allow water to flow from one pit to another as shown in the Fig. 15.1. In each pit a rectangular cloth tank, 6′ × 3′ × 1½′ in size, made up of coarse cloth should be tied to the bamboo poles fixed in the pit. A similar open tank (5′ × 2½′ × 1′) is made up of round meshed mosquito curtain cloth, projecting 3-4′ only out of the water surface.

Small hatching pits without the cloth tanks hapas are commonly used near the bundh type tanks in Midnapore (Sinha *et al.*, 1973; Singh *et al.*, 1979).

Nursery ponds. They may be perennial or seasonal, retaining water for a long period. The advanced fry or early fingerlings should be kept in such ponds for 2 to 3 months so that they attain advanced fingerlings size (4″–6″ in length). Depth of water should not exceed 6 feet. Scope for easy netting should be available for which purpose weeds must be less. An alternative device is that paddy fields holding water to a depth of 1½ ft to 2 ft for 2 to 3 months may also be useful for rearing early fingerlings.

Stocking ponds. Giving consideration to requirement of easy netting facilities and availability of a long run for the fish, perennial ponds, 1/3 of an acre or larger in area and over 6 feet in depth, are ideal stocking ponds. A scene of stocking pond is evident from figure given before (Fig. 12.1).

Breeding ponds. Minor irrigation tanks (perennial or seasonal) traditional ponds and bundh type tanks are most important sites as the majority of Indian carps breed in them. When the bundh is full of water, the same flows over and submerges the adjoining shallow areas or fields. Breeders are reared there or they are introduced from elsewhere into the pond. It has been observed that during peak of monsoon, the fish become more active and enjoy moving up to the shallow inundated grounds or field, which usually contain 1 or 2 ft deep water. The move of the matured fish is mainly for the purpose of spawning. The irrigation tanks of South India and Madhya Pradesh are functionally like the bundh type tanks and large ponds of North-east India. It is an established fact that the common carp, *Cyprinus carpio,* breed in the ordinary ponds, though the fish is an exotic major carp. Lakshmanan *et al.* (1971) and Singh *et al.* (1972) have given useful information about the composite culture of the exotic carps. Relevant informations with illustrations have been provided before in Chapter 5, namely, 'Induced Breeding'.

Procurement of seeds. Method for collection of seed varies slightly in different localities but in the rural areas of Eastern India, the collection is done in the following way :

A piece of mosquito net with round meshwork is laced to the small bamboo poles at either end, held by two persons in a particular position against the water current. At a gap of every 10-15 minutes or so, the net is lifted up and the entrapped eggs are collected and gently transferred to an earthen 'hundi'. Two persons are required for operation of a net with the help of a boat. In some places, bamboo poles are permanently fixed at different positions, suiting to the possibility of collection.

The fry of the major carp is collected in plenty from the rivers and rivulets of West Bengal, Bihar, Orissa and Uttar Pradesh. Kolkata (Howrah), Patna, Allahabad are the main centres from where carp fry and fingerlings are supplied to various places where these are not locally available.

In the Gangetic water area of Patna, the carp fry is collected with a special type of closed meshed funnel shaped net with a tail part formed of cloth, called 'Gamcha'. The narrow tapering portion is fitted with a ring. The net is laced to bamboo poles with its mouth facing the current of water in the shallow areas. After sometime, carp fry collects in the 'Gamcha' and the same is transferred to hapa or to a small pit (a modified form of hapa) which is supported on poles with proper arrangement for the circulation of water. Figures given in Chapter 5 illustrate the account given here for procurement of seeds.

In West Bengal, the collection of fry is transferred to earthen hundies which are to be given shake regularly during the transit so as to keep the contained water of hundies aerated. In Madras (now Chennai) and Kolkata generally tin containers are used for this purpose. The Central Inland Fisheries Research Station,

Barrackpore (West Bengal), has designed special type of containers in which there is provision of oxygen in sealed containers. By using these containers, transport of fingerlings to a destination is safely done. Additional informations have been given in chapter 7, entitled 'Transportation of Seeds'.

Manuring of Nursery Ponds

Before transfer of the seeds to nursery ponds, manuring of the latter is required to boost up natural food production. The manures used are either inorganic substances, such as chemical fertilizers, superphosphate, ammonium sulphate, sodium nitrate or organic substances such as cow-dung and house-sweep, etc. The use of inorganic fertilizers leads to manifestation of microscopic floating algae, whereas organic manures enhance zooplankton population. As zooplanktons are relatively more valuable food for early carp than the floating algal matters, it is advisable to use organic fertilizers. In fact, the use of raw dung manifests production of planktons within a short time. A simultaneous addition of green manure such as leaves of plants, like lantana, and raw cow-dung boosts up production of plankton in water.

Transfer of Eggs to Hatching Pits

Collection of fish seeds should be thoroughly examined in shallow flat dishes to locate the presence of any insect or debris as impurity and discard them. They should be transferred to a sieve with a meshwork of 3.0 × 3.5 millimeters in diameter, kept partially immersed in water, to separate only viable and healthy seeds (fertilized eggs). As the fertilized eggs of major carps are larger than the sieve space, the separation is easily operational.

The eggs are now transferred to the inner tank or hapa made of round meshed mosquito netting. The Calico Co. net is considered as ideal for this purpose. The eggs normally hatch out 2.5 hours after collection, but as the eggs are in different stages of development, the time taken by eggs for hatching out may not be the same. The hatchings pass easily through the meshes of hapa into the outer cloth tank, as evident in figure, but the burst egg membranes are left over in the inner net (Fig. 15.2). These are removed after all of the eggs are hatched.

Nursing of the Fry (Hatchings and Larvae)

The hatchlings (4 to 5.5 mm in length) are, on the 3rd day after hatching of eggs, collected in hundies and transferred to nursery ponds (Fig. 10.3). The tiny hatchlings or fry have to sustain several kinds of constraints in the new environment (Fig. 15.3). The greatest hurdle is the insect population, many kinds of which are predaceous. The remedial measure on this account is the chemical treatment with insecticides. This will have positive effect and occurrence of mortality of the hatchling or fry will be reduced. The next important requirement is the quality of water, which is to be properly examined and ideal condition ascertained and further next requirement is the provision of feed from external source. The results of experiments at different scientifically managed fish farms have given indication that maximum growth and highest rate of survival of fry are obtained when they are fed with zooplanktons, preferably water fleas. The growth and survival of fry is considerably low on providing only algal food. Artificial feeds like rice bran, powdered groundnut (these two in the ratio of 1 : 1 by weight), coconut and mustard cake may also be used as suitable feeds. Feeding of the grass carp with hydrilla, kept in bamboo platform in the centre of pond, gives excellent growth to this fish. If the nursery pond is kept free of predaceous insects, and abundance of food (mainly of zooplankton) is provided, the rate of survival of the fry may reach 90% level.

Rearing in ponds. On giving suitable conditions, the fry grows fastly and the food requirements increase endlessly. The advanced fry and early fingerling are, therefore, collected from nursery ponds and released in the rearing ponds where larger areas and new overall conditions suit their requirements. They are kept there till they grow further to attain the advanced fingerling size. Provision of food from external source, *i.e.,* supplementary feeds, and insect-free conditions are maintained for the development in nursery pond.

Transfer to stocking pond. Only the young ones in advanced fingerling forms are sorted out from the stock of collection from the rearing pond, obtained by simple nets operation and released in the nearby located stocking ponds. The density of stocking should be in accordance with the size of pond, its food resources, natural fertility, and species of fishes already available in the stocking pond. If all kinds of fingerlings of major carps are available in the stocking collections, the combination should, as per advice of fishery experts, be as given below :

(i) For Catla, rohu, and mrigals, the ideal ratio of combination is 4:3:3; the stocking rate should be 3750/ha.

(ii) For composite culture of six major carps, *i.e.*, catla, Rohu, mrigal, silver carp, grass carp and common carp, the ideal ratio of combination is 2.5:5:2, 5:2:2:5. The stocking rate, in this combination, should be 5,000/ha.

(iii) If combination of seven major carps is desired, then the ratio should be 2.5:5:2.5:2:2.5:0.5, the last one being for *L. calbasu.*

The best time for stocking is April and May. The stocking density can be increased by enhancing the nutrient and food conditions proportionally, that is, by providing proper fertilizers and supplementary feeds from external sources. As mentioned before also, a provision of hydrilla kept in a bamboo platform in the centre of the pond give excellent growth rate of the grass carp (Fig. 3.5). In the periodic addition of hydrilla, the amount should be increased gradually so as to suffice the need of growing grass carp. Generally, after 6 months' time, the grass carp picks up 2.5 kg body weight.

Post Stocking Management

After stocking advanced fingerlings in the stocking ponds, the foremost attention should be given to the supplementary feeding. Artificial feeding is done with the carbohydrate rich cereals waste in association with the vegetable oil cakes which contain fairly rich protein content. A mixture of rice or wheat bran and mustard or groundnut oil cake in the proportion of 1:1 serves the purpose. The amount of feed should ideally be 2.5-10% of the body weight of the fish stocked and suitable time of application of feed is early morning. In the case of grass carp, mixing of the chopped piece of green vegetation along with the feed serves the purpose of yielding good result. If it is marked that the stocked fish consume all the feed provided, and no waste is left, the feeding schedule should be gradually increased.

After the pond is stocked, a periodical check up is necessary. To check the growth of the stocked fish, it is desirable that there should be netting operation, as a sample work, once a month or once in two months. Giving weightage to the performance of fish in the sense of growth rate, the schedule of external fertilization or addition of supplementary feeding or both may be occasionally altered. Due to differential growth rate, the size attained by the stocked major carp may be widely varied. In such a situation, larger sized fishes are separated by netting operation and transferred to another stocking pond. In this way, the smaller major carps get a chance to flourish, *i.e.*, grow at faster rate.

Harvesting. Harvesting is done after 6 months, specially for grass carp and after one year periodically for all the major carps. The best time of harvesting is summer when the water level is low and netting operation is easy. Netting, that is, the cast net operation is properly done in the early morning, as the water column is calm at that time but it gets heated up with the rising sun.

16

Hilsa Culture

Economic value of hilsa fish in eastern India, particularly West Bengal, is more than that of Indian major carps, as the price value is always higher and demand also is not the less. Hilsa reigns the minds of Bengalees and it finds references in poem, literature, rhymes, religious stories. Nevertheless, on some festive and religious occasions a menu of some preparation of hilsa is a must. As the demand of hilsa is invariably high and supply in the market is less, craze of hilsa is always on top. This situation has led to smuggling activities in border towns of Bangladesh. Nowadays, a good sized hilsa is priced 50% to 75% higher to rohu.

The nutritive value of hilsa keeps the fish at superior level to other fish like major carps. 100 gm of hilsa usually contains 22.0 gm of protein, 19.5 gm of fat, 180 mg of calcium, 250 mg of phosphorus along with other nutrients. Hilsa from Mahanadi river has the highest fat content (20%) and those from Narbada the lowest. Accompanied with superior nutritive value, the flavour of hilsa flesh is of high order. In earlier days, the possibility of culture of hilsa was not within imagination, but the situation has changed rapidly and technique has been developed to tame this fish in controlled conditions (Kulkarni, 1950; Jones and Menon, 1951; Bhanot, 1973; Malhotra *et al.*, 1973).

Distribution in Natural Conditions

Basically hilsa is a marine fish, being found in abundance in the Brackish waters of India, such as the brackish waters of Kerala coast, Chilka lake and Hooghly estuary. Hilsa migrates to freshwaters and is caught from rivers of Ganga, Brahmaputra, Padma, Mahanadi, Daya, Godavari, Krishna, Cauvery on the east coast and Narbada, Tapti, Purna Ulhas, Kali on the west coast.

A few species of hilsa viz. *Hilsa toli, H. Kanagurta, H. ilisha* are available along the coasts of Indian sub-continent. Out of these, *Hilsa ilisha* is the only species that migrates deep into inland waters for breeding and feeding and, in this pursuit, it is greatly acclimatized to the fresh water conditions. In the west coast, species of hilsa are distributed on the Saurashtra coast and freshwater areas on the mouth of rivers Narbada and Ulhas. However, its distribution is conspicuous also in the Palk Bay foreshore areas at the mouth of Godavari, Tamil Nadu Coast, Visakhapatnam coast, Orissa coast (in abundance in Chilka Lake) and Bengal coast (particularly Hooghly estuary).

Species of hilsa grow to 30 cm size in the case of male and 35 cm in female, when mature (Fig. 16.1). These sizes are attained generally after one and half years of age. The life span is generally from 5 to 7 years. Hilsa is caught in large numbers in West Bengal and Orissa from the lower regions of nearly all rivers on the east coast, and in large numbers from the Chilka lake and Matlah estuary. The main catches are after the monsoons from these flooded estuaries of the Ganges and Mahanadi where fish migrates in large numbers. However, some fishing is done also during monsoon and winter months. Various kinds of fishing gears are used to catch hilsa but 'Sangha Jal', gill nets and bag nets are more popular.

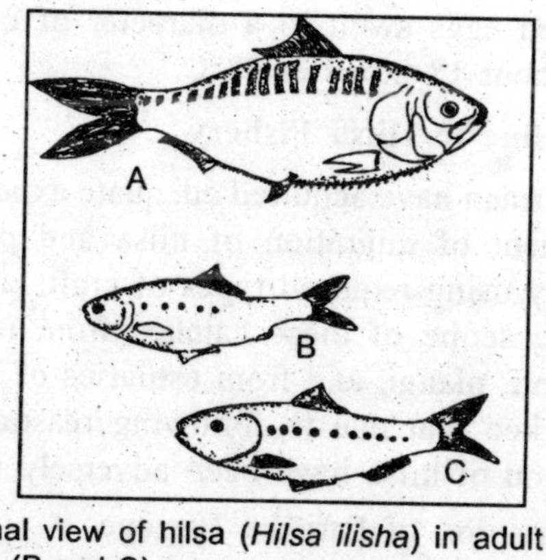

Fig. 16.1. External view of hilsa (*Hilsa ilisha*) in adult form (A) and two young forms (B and C).

Migratory features. Opinion varies in respect of its time of migration, age, and season of migration, choiced food and feeding habits. Hilsa is variously described *e.g.* as an anadromous fish, as diadromous fish, a fish of estuarine areas, coastal areas, near shore areas, etc. Several biologists including Kulkarni (1950), Karamchandani (1961) and Bhanot (1973) worked on various biological and cultural aspects of this prized fish of India. Causative factors of migration have been attributed to change of temperature, change of season, turbidity, current, age, etc. Two types of migration in Bhagirathi, *viz.* surface migration, while ascending the rivers with developed gonads, and bottom migration which means descending to estuaries by the spent fishes have been reported.

Food. Planktonic food such as diatoms, algae, volvox and microphytes have been found to form main food of ascending fishes, while good quality of bottom muck along with above mentioned planktonic food have been found to be the main constituents of food of the returning fishes. In brief, hilsa is a strict plankton feeder. Its gills and gill rakers are modified into a very fine sieve that must filter all microscopic food matters entering the buccal cavity with respiratory water current.

Fecundity. Like all clupeid fishes, *Hilsa ilisha* has very high fecundity; *i.e.,* above 7 lakh eggs/kg body weight. Unlike I.M.C. hilsa matures only at +1 year of age. Eggs are laid in calm water. They are desmersal but may be buoyed up and drifted away. The ripe water soaked eggs swell to a diameter of 2 mm Hatching takes place in about 18 hours at 28°C.

Reason for Decline in Hilsa Fishery

Bengal fishermen have acquired adequate experience to sense the commencement of migration of hilsa and proper time for catching them by using required types of crafts and gears. They have an alluring scope of these catches from river Bhagirath, Rupnarayan, kunti, plangi, and from estuaries of Bay of Bengal. It has been marked that due to following reasons migration as well as population of hilsa have been adversely affected :

(i) Construction of Farakka Barrage.

(ii) Reduction in water discharge.

(iii) Excessive discharge of pollutants.

(iv) Excess fishing actions of grown up as well as juveniles of hilsa.

Also the use of pesticides on the agricultural crops to control pests has led to access of the pollutants to the aforesaid rivers and this has led to deterioration of the water quality. This factor is also responsible to some extent for the population and prevalent trend of migration of this fish to deep inland water areas.

Total hilsa landings (all species of hilsa) at national level during 1980-85 were 6996, 4507, 3618, 3676, 5897 and 9154 tonnes respectively against 2683, 1128, 1011, 625 and 3294 tonnes in West Bengal.

Present Status of Hilsa Population

Migration as well as population of hilsa have became grossly affected mainly due to construction of Farakka barrage. However, the extent of damage caused is yet to be explored. Political barrier between India and Bangladesh has obviously no effect on the migratory fish, hilsa, which migrates to far off destination, if any physical barrier is not in-between. Unfortunately current data on total catch per unit effort, landings, length-weight relationship of the hilsa and related data from study of population dynamics and assessment of available stock of fish from both countries are not available. It can be concluded that there is not yet any effective mechanism to obtain aforesaid and other relevant informations.

From the record of marine fish landings during 1991-94, the extracted information on hilsa is shown in Table 16.1.

Hilsa Culture

Sporadic experiments on hilsa rearing in impounded waters and artificial feeding by stripping of broad fish collected from natural sources have been made several times at several places *viz.* Allahabad, Puri, Chilka, Barrackpore etc. and the records were found to be successful. Hilsa has been found in catches landed from bheels of Assam, Gumti Reservoir in Tripura, and Matalila Reservoir. It has established itself in Ukai reservoir of Gujarat. Although there is absence of gravid and spent brood in

Table 16.1

Hilsa Fish Landing during 1991-94 (in Tonnes)

States/UTs	1991	1992	1993	1994	Remark
Andhra Pradesh	483	483	1,638	0	
Orissa	5,561	8,174	6,824	5,189	Next highest
Tamil Nadu	519	1,177	635	1,405	
West Bengal	23,534	23,534	23,534	27,920	Highest
Andaman	1,233	3,141	1,460	964	
Pondicherry	146	196	629	0	
Gujarat	1,251	1,759	1,588	2,138	
Karnataka	–	–	0	0	Nil
Kerala	–	–	0	83	
Maharashtra	1,312	1,511	1,190	769	
Goa	162	162	162	162	
Lakshadweep	–	–	0	0	Nil
Daman and Diu	153	111	97	80	
Total	35,254	40,248	37,757	38,710	

Source : Handbook of Fisheries Statistics, 1998, Govt. of India.

Ukai reservoir, hilsa collection indicates spawning in this reservoir. This fact is a controversial one. Examples cited indicate that hilsa can thrive and grow in fresh impounded waters.

A case history of hilsa rearing in West Bengal is indicative of successful hilsa culture (Fig. 16.2). Hora (1941) and Bhanot (1973) have given valuable informations. A freshwater fish farm of about 3 ha having a water spread area of about 1 ha is situated in the Aschitpur Mauza of Dumudah village under P.S. Balagarh, dist. Hooghly of West Bengal has been successfully used for hilsa culture. The place gets connected with Balagarh river during monsoon months and is subjected to low and high tides during full moon and new moon. Water flow in the farm is regulated through a sluice gate during January to May and again from September to October. After intake of water, the tank is manured with mahua oil cake at the rate of 2500 kg/ha and lime 280 kg/ha.

Cattle dung is not used. With the commencement of induced breeding, which comes generally in May, the ponds are stocked with IMC spawn for harvesting after 15-20 days. Rearing of spawn to fry continues till last week of September. In the in-between period of 2 months the farm is manured with mustard oil cake.

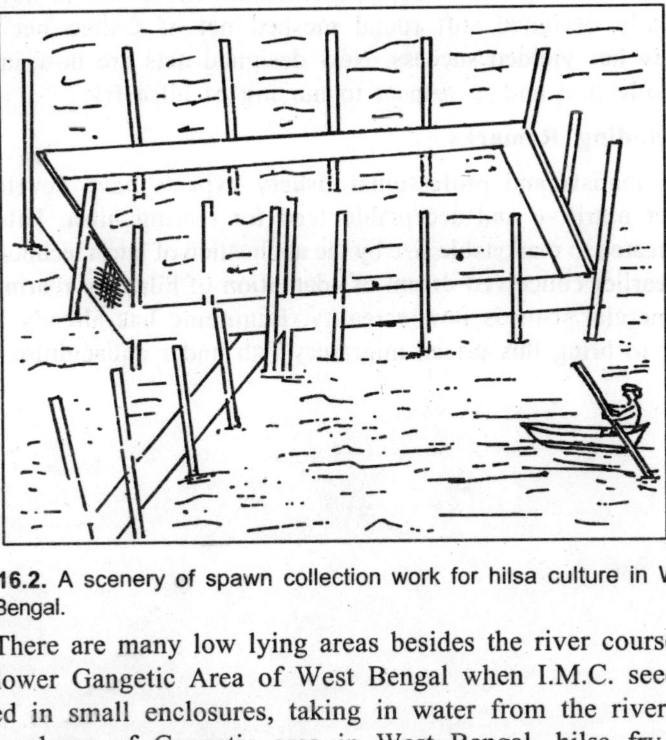

Fig. 16.2. A scenery of spawn collection work for hilsa culture in West Bengal.

There are many low lying areas besides the river course in the lower Gangetic Area of West Bengal when I.M.C. seed is reared in small enclosures, taking in water from the river. In many places of Gangetic area in West Bengal, hilsa fry are available in plenty. In the fish markets of Murshidabad, Berhampore, Lagola and Farakka, Dhutia and Jangoepur there is a regular supply and sale-business of hilsa fry.

Constrains in Rearing Hilsa

One of the main constrains in rearing of hilsa fry is transportation of fry from the collection centre to the ponds. It is a noticeable fact that hilsa is sensitive to oxygen deficiency and handling. It has been found that mere 30% mortality in 2.5 hour duration occurred on experimental transportation of 1,00,000 spawn under oxygen pressure in 18 litre of water. On the other hand, transportation of 75,000 seed, slightly larger in size, packed in 18 litre of water with oxygen showed negligible mortality after the same period of time (2.5 hours). It has been found that maximum mortality of fry occurs during netting by traditional nets, such as gill net and drag net. Recent experimental endeavour to develop specially designed soft round meshed net of Calico net type variety has yielded success. New designed nets are now in use for collection and in respect to nursing of hilsa fry.

Concluding Remarks

Scientists and professional fishery expects have developed proper nutritive and acceptable feed for rearing hilsa. Hilsa is now reared to marketable size by the application of latest technology. The earlier conceived dream of adaptation of hilsa fry rearing on commercial scale is now a reality. Beginning has already been made to bring this prized migratory fish under aquaculture.

17

Trout Culture

Trout and Salmon represent popular game fishes of the world and, because of palatable flesh, are two important fresh water fishes. They are also well-known as cold water fishes. Their proper growth and survival are dependent on two main factors, namely (1) an abundant and continuous flow of clear cold water, preferably in running streams, (2) availability of adequate amount of food *i.e.* artificial or supplementary feeds. From the trout culture point of view, the Himalayan ranges can be divided into three zones, *viz*. Western, Central and Eastern zones, comprising Kashmir and Ladakh in the west, Kumaon hills and Kullu valley in the centre and North East Frontier Agency in the East. The division of these sectors is based on the water temperature of these regions which is congenial for trout culture and is well within the limits (25°C).

The various species of trout that are well suited for culture in India are the landlocked varieties of brown trout, *Salmo trutta,* the rainbow trout, *Salmo girdneri* and the brook trout, *Salvelinus fontinalis.*

The brown trout. They are native of European waters and introduced from there to various countries of the world. In India the brown trout was introduced into Kashmir in 1901 by the kingdom and is flourishing in several snow fed streams of Kashmir since then (Fig. 17.1). They are well suited for culturing in cold water regions where temperature does not exceed 20°C. The quality of fish as human food is slightly less valued than of water trouts.

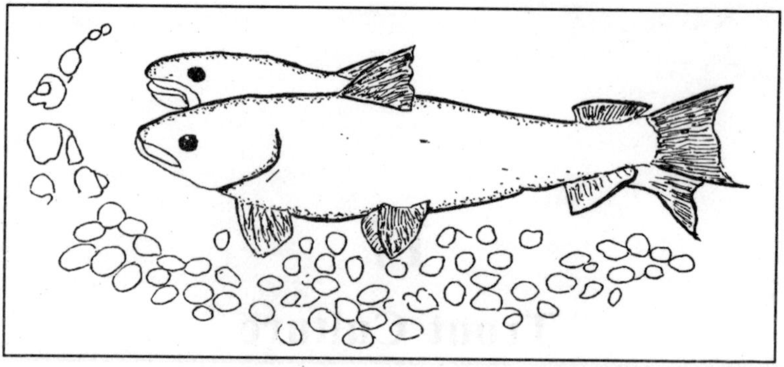

Fig. 17.1. Outline view of brown trout, *Salmo trutta*.

The rainbow trout. This variety is most popular of all trouts. They are native of American waters and have been introduced as game fish throughout the world (Fig. 17.2). They have greater adaptability to temperature, water quality and water quantity. They can withstand higher temperature upto 25°C. They have high stocking density, moderate growth rate and are more resistant to diseases.

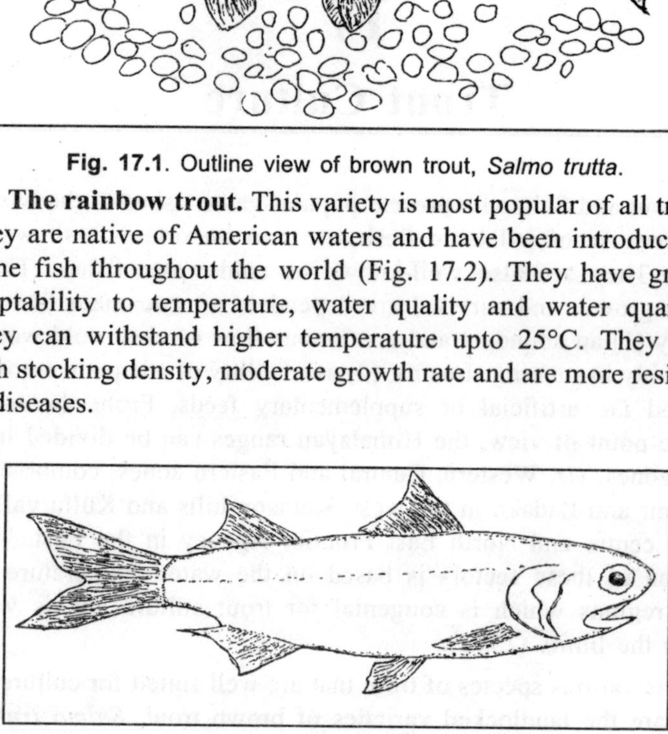

Fig 17.2. Outline view of rainbow trout, *Salmo gairdneri*.

The brook trout. This fish, represented by *Salvelinus fontinalis*, was very popular when introduced in Indian waters, in 1919 and later transplanted from Kashmir to Kullu waters of Himachal Pradesh, but is now losing status of high demand because of poor adaptability (Fig. 17.3). Their transplantation in Bhowate Naini Tal and several snow fed streams of Kumaon did not yield success in the long run. It is rather difficult to culture this fish because of poor tolerance to temperature (about 18°C). Secondly, this fish has the disadvantage of susceptibility to diseases (fin-rot, furunculosis) and also to pollutants or toxicants. However, a superb quality is that this fish has fatest growth rate among the three trouts in question and the quality of its flesh is most palatable.

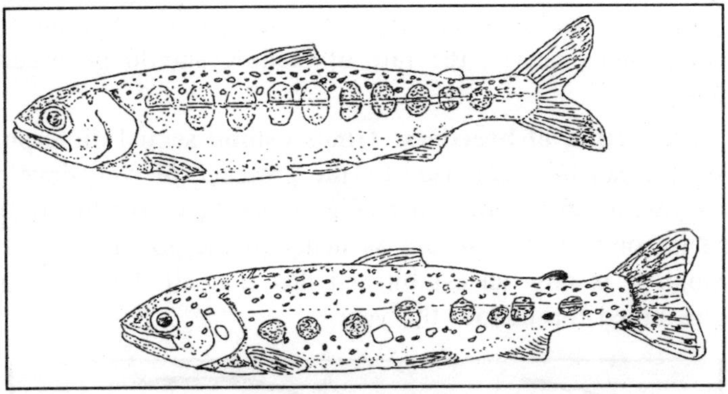

Fig. 17.3. Outline view of brook trout *salvelinus fontinalis.*

Management in Culture of Trout

The basic requirements in the culture practices of trouts are as follows :

1. After experimentation, expert advice for culture practices with trouts is that trouts belonging to age groups of 2 years alone should be selected for stocking in tanks, since they (females) possess quality eggs. Females and males are stocked in separate tanks of 10 × 3 × 1.5 m dimension, constructed with a gradient of 1 m 200 with cemented sides and cemented or gravelly bottom. The depth of water should be maintained at 1.0 m at the fore end.

2. Stocking and artificial feeding of breeders. Adults in the range of 30-40 cm in size and 400-500 g in weight are selected on visible health criteria and stocked at the rate of 5 fish/m^3. The total stocking density comes to be 150 breeders for each tank.

Artificial feeds of various kinds have been tried and results compared for excellence (Phillips, 1929; Sehgal *et al.*, 1984). One of the suggestions is that artificial feeds prepared with items of animal and plant matters, mixed with mineral mixture in the proportion of 65, 30 and 5 respectively enriched with vitamin are made into pellets (8-10 mm in diameter and 12-15 mm in length) and fed twice daily at 2–3 c/r of body weight. Another suggestion is that half boiled fish, preferably of *Schizothorax* spp. or common carp (*Cyprinus carpio)*, can be used as suitable feed for breeders. The amount to be introduced in water should be 5% of body weight twice daily. It has been noted that mixing of some salt with the feed yields good result. As the date of stripping of the breeder comes nearer, the rate of feeding should be gradually reduced.

3. Selection of breeders. Trouts exhibit sexual dimorphism; the males can be recognised by the presence of well developed lower jaw, as shown in figure (Fig. 17.4). Selective breeding on merit is conducted to secure as many of the above qualities as possible. It is now an established fact that all the three major species breed throughout the year.

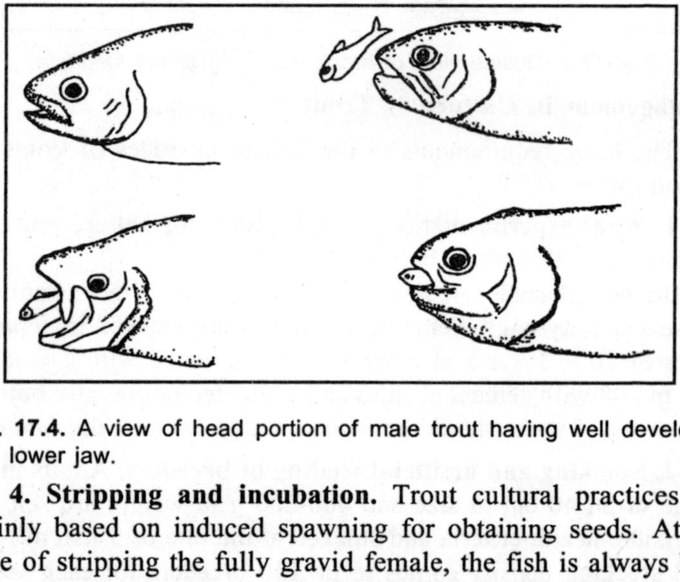

Fig. 17.4. A view of head portion of male trout having well developed lower jaw.

4. Stripping and incubation. Trout cultural practices are mainly based on induced spawning for obtaining seeds. At the time of stripping the fully gravid female, the fish is always held upwards and stripping is done by employing gentle pressure on the sides of the abdomen with the use of hand. The eggs are restripped into bowls. First the eggs from the female and then milt from the male are collected in the same bowl or receptacle which contain water (if wet method is employed) or may not contain water (called dry method). Dead eggs, recognized by floating condition are removed.

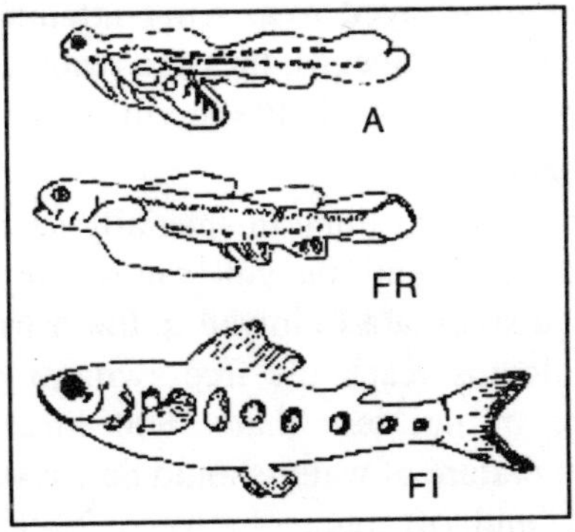

Fig. 17.5. Outline view of Alevin (A), Fry (FR) and Fingerling (FI) of salmon.

Fertilized eggs are allowed incubation in specially designed tray or jar available until hatching in water preferably at about 10°C with dissolved oxygen concentration at 7 ppm. During incubation a constant desired flow of water is maintained keeping a check on the escape and mortality of eggs. The dead eggs should be carefully removed everyday, whereas the green eggs should not be disturbed, as they are viable. The green eggs become eyed within a period of 40-50 days at a water temperature of 8-10°C in the case of brown trouts and within 20-30 days in the case of rainbow trout at the same water temperature, whereas brook trouts require a longer period than both. In general, hatching may occur in 30 to 45 days at 10°C.

Fig. 17.6. Collection of trout seeds from a stream.

6. Transhipment of eyed ova. After attainment of the eyed condition by the developing eggs, they are ready for transhipment, preferably in plywood bags (45×40×40 cm) which can hold about 20 incubation trays.

7. Rearing of alevins is done in specially designed troughs of aluminium or in concrete until the yolk sac is completely absorbed. Conditions of incubation are followed in the rearing of alevins as well. When the alevins reach the free swimming state, flow of water at the rate of not less than 3600 litres per minute is required. The temperature of water should be constantly isothermal between 10-18°C and pH range between 70-85 and dissolved oxygen concentration at 5 ppm. These conditions are not easy to achieve and these ecological limitations have made trout culture an uncertain practice in many places of the world. Only few streams throughout the world meet with such conditions where rearing of alevins is a flourishing cultural practice.

Nowadays several trout hatcheria have been planted with provision of all ecological essentialities in controlled conditions. These are located at Avalanche in the Nilgiris, Rajamally in Kerala, Laribal and Harwan in Kashmir, Kullu and Barot in Himachal Pradesh. As such, the problem of seed procurement has been solved to a great extent. The fries are generally fed on the bound diet in the form of crimps, to begin with, and later (for fingerlings) switched on to pellets prepared by mixing the new raw materials with water, making it a dough.

8. Rearing of fingerling is preferably done in natural springs. Artificial ponds and raceways are also used in several places. The raceways of 25 × 5 × 1 m dimension, with a bottom slope of 1-3% of the depth, have been found by Jhingran (1968), Sehgal (1979) very ideal for the growth and health of the growing fingerlings. The long run in the raceways provides required opportunity of exercise to the young ones. During construction of the raceways, water inlet is constructed in such a form that a virtual fall of at least a few inches is provided (Sehgal, 1976; Jhingran and Sehgal, 1978).

There is precised system for regulating flow of water in the raceways. The criteria of release of flow of water depends upon the age of the fingerling. For tiny fingerlings, less flow of water is required, and in the grown up fingerlings, the requirement increases. For 1000 fingerlings, a flow of 1 litre water per minute, uninterrupted for a month is sufficient. As metabolic waste products may accumulate to invite diseases and depletion of dissolved oxygen concentration are liable to occur, the regularity in flush of water suffices the purpose.

Fig. 17.7. A scenery of rearing of young one of trout in circular pools.

While introducing fingerlings in the raceways or artificial ponds or natural springs, another requirement of management is the grading. Box-graders are used for this purpose and selected grown up fingerlings have uniform growth and, on harvesting, fish yields of uniform size are obtained. In the view of Sehgal (1979, 1984) this uniformity allures marketing prospects.

Some Common Diseases and their Control

Various species of trout that are in use for culture practices are the landlocked varieties of brown trout, rainbow trout and brook trout. These fish prefer to live in clear running water with stream like conditions. However, they are not immune from infection by bacterial, fungal, parasite diseases. In Indian conditions, Ramakrishna (1976) has identified five kinds of diseases in trouts.

The types of diseases, main symptoms and remedial measures are given here :

Disease	Symptoms	Remedial Measures
1. Whirling	Leaping and screw type of movement.	The diseased trout should be isolated and eliminated.
2. Pin head	Big head, and emanciated body	Mix calomel at 0.5-1.0% with feed.
3. Fin rot	Fin tips being eaten up	Give copper sulphate bath in the dose of 1:2,000.
4. White spot	Appearance of white spots on fins, gills and body	Dip in formaldehyde solution (1:4,000) for 15 minutes.
5. Fungal attack	Presence of white patches on the body	Give treatment of Malachite green solution (1:20,000) for 1 minute.

It has been marked by culture practices that the specimens of brown trout are more resistant to whirling diseases and those of rainbow trout are resistant to all kinds of diseases. However, the specimens of brook trout are less resistant to various kinds of diseases, pollutants as well as toxicants.

18

Live Fish Culture

Culture of air-breathing fishes, in short, also known as 'Live Fish Culture', is an age old practice in India. It is popular in Japan, Southeast Asian countries, particularly Thailand, and also in India. In wild condition it is found in abundance in all kinds of low lying land areas where is accumulation of water for great part of the year, if not whole year, due to heavy rainfall. The water bodies are generally unconnected from the river or running stream of water. Due to derelict situation, typical water-logging symptoms may develop, such as infestation with various types of weeds or aquatic vegetation and the whole water area may turn swampy. Also bheel, and Jheels in some parts of India become heavily infested with aquatic vegetation, get chocked and, due to resultant eutrophication, get converted to swamps. These water bodies harbour several such organisms which form food for fishes. They include phytoplankton (microphytes, macrophytes) zooplankton, benthic organisms (worms, aquatic insects or their larvel forms) and plenty of decaying organic matters. In such an congenial environment, fishes which can breath by gills or take in atmospheric oxygen through the accessary respiratory organs find paradise to thrive and pass their life from egg to adult form without any aid of supplementary feeds. The fishes are easily caught by traps or simple gears with net meshwork. Whatsoever, the live fishes are important food fishes for human consumption, carry commercial value and are suitable for culture in confined water. As they breed in ponds, their culture practices is easy. However, because of carnivorous habits, they cannot be cultured along with carps.

Major constraints in the practice of live fish culture is poor survival of the fry and fingerlings, mainly on account of cannibalistic habits of these fishes.

Culture Practice for Live Fish

The common live fishes of commercial value in India are the following species :

Common Name	Family	Genus/Species
Murrels		
1. Giant murrel	*Channidae*	*Channa marulius*
2. Stripped murrel	"	*Channa striatus*
3. Spotted murrel	"	*Channa punctatus*
Catfishes		
4. Magur	*Claridae*	*Clarias batrachus*
5. Singhi	*Saccobranchidae*	*Saccobranchus fossilis*
Perch		
6. Climbing perch	*Anabantidae*	*Anabas testudineus*
7. Giant gourami	*Osphronemidae*	*Osphronemus goramy*

Murrels

They belong to the family *Channidae* (formerly called *Ophicephalidae*), as mentioned before. They have marketing prospects because of good flavour of flesh and have availability throughout the year at considerably cheaper rate, particularly in North-east India. Four common species of *Channa* (popularly known as murrels) are *Channa marulius, C. striatus, C. punctatus* and *C. gachua.* Morphologically, they are characterised by the following features of fins and lateral line :

1. *C. marulius* : D.46; p.18; V.6; A.32; L.l. 65-66; L. tr. 6/11.
2. *C. striatus* : D.42-43; p. 16-18; V.6; A.24-25; L.l. 58-60; L. tr. 7/10.

C. punctatus : D.29-30; p. 16-17; V.6; A.20-22; C.12; L.l. 35-37; L. tr. 4-5/7.

C. gachua : D. 35-37; p.15-16; V.6; A.21-2s; C.12; L.l. 41-45; L. tr. 4/7.

C. marulius attains maximum length of 70-75 cm and weight of 2.5-3 kg when fully grown (Fig. 18.1). However maximum length of *C. striatus, C. punctatus* and *C. gachua* are respectively 28.5 cm, 15.3 cm and 13 cm respectively (Figs. 18.2, 18.3, 18.4). They are hardy fish with less mortality rate and oftenly cultured in combination with *Saccobranchus fossilis* and *Anabas testudineus.* However, these fishes are also suitable for monoculture practices. In rural areas, particularly in Eastern India, they are invariably found in ponds, and small rivers. Specimens of *C. striatus* and *C. gachua* are also found in paddy fields and other shallow water areas. Murrel, have distribution also in Pakistan, Bangladesh, Burma and Ceylon (Sri Lanka).

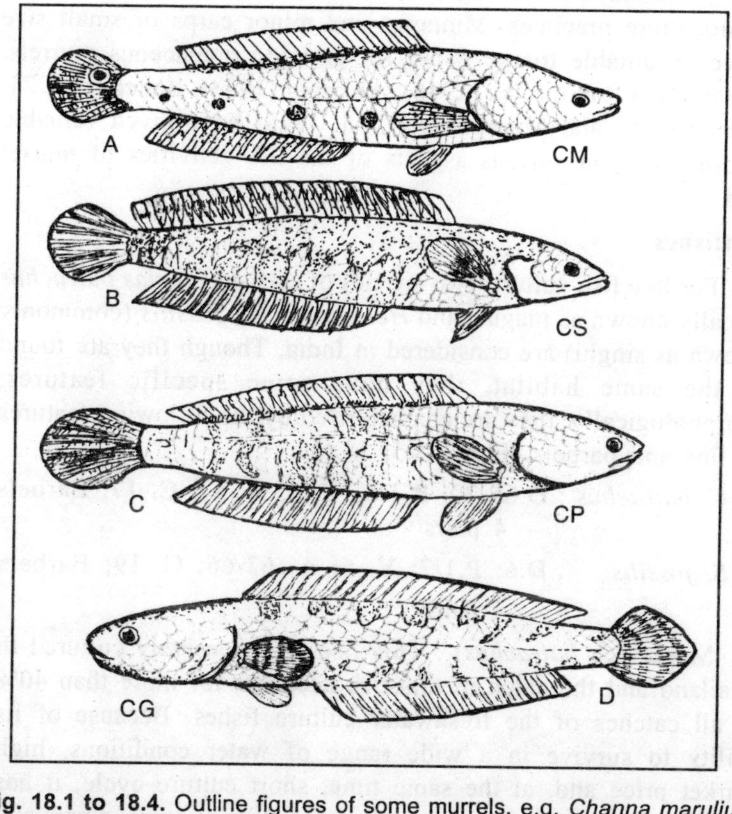

Fig. 18.1 to 18.4. Outline figures of some murrels, e.g. *Channa marulius* (CM), *C. striatus* (CS), *C. Punctatus*, (CP) and *C. gachua* (CG).

Murrels in general are pond breeders and young ones (in shoal) are given parental care for several days. They breed throughout the year, but the peak period extends from April to June. The young fishes feed on small crustaceans and vegetations, but adults are highly predaceous (carnivorous, piscivorous). They attack other fishes and consume them as food. Experiments have shown that out of many fishes, *puntius sophore* and *Oxygaster bacaila* are suitable forage species for murrels. However, for *C. striatus,* tilapia serves as a forage fish.

Recent research pursuit has shown that all the three species of murrels can be bred by pituitary injection. The seeds thus obtained are transferred to nursery ponds, then to rearing ponds and the product, fingerlings, are introduced to stocking ponds for monoculture practices. Minnows and minor carps of small size serve as suitable forage fishes for growing predaceous murrels. Hosaini and Rahimullah (1946), Alikunhi (1953), Banerjee, 1974, Parmeswaran and Murugesan (1975, 1976) have given valuable informations on various aspects of the life activities of murrel fish.

Catfishes

For live fish culture, two species of catfish, *Clarias batrachus* (locally known as magur) and *Heteropneustes fossilis* (commonly known as singhi) are considered in India. Though they are found in the same habitat, they have some specific features. Morphologically, they are characterised by the following features of fins and barbels (Figs. 18.5, 18.6) :

C. batrachus : D.65-70; P 1/8; V. 6; A. 47; C. 17; Barbels — 4 pairs

S. fossilis : D.6; P.1/7; V. 6; A. 62-66; C. 19; Barbels — 4 pairs

Magur (*C. batrachus*) : This fish is extensively cultured in Thailand and the catch of this fish accounts for more than 40% of all catches of the freshwater culture fishes. Because of its ability to survive in a wide range of water conditions, high market price and, at the same time, short culture cycle, it has gained popularity in India as well. It inhabits swamps, rivers and ponds and feeds on insect larvae, crustaceans, worms, tiny fish and debris. It is basically carnivorous in feeding habits, but it has neither predatory nor habit of cannibalism. However, the fingerlings show cannibalism.

Young one of 5 to 10 cm size are collected from natural source of water by rearing the fish in ponds in controlled conditions and obtaining, seeds through induced breeding. They are stocked in pond at the rate of 50,000 fingerlings/ha and when cultured with carps the stocking rate may be reduced to 25,000 fingerlings/ha.

During culture period, the fish is provided feed in the combination of cheap dried marine trash fish and rice bran, mixed together at different proportions. The growth of magur is rapid but the rate depends on feeding. If the water quality in the pond is good, a production of 4,000 kg/ha in six months can be achieved.

The fish breeds during monsoon in confined water and is known to lay eggs in the mud holes on the bank of pond. Fish seeds have been raised by induced breeding technique, known as hypophysation, pioneer efforts were made by Ramaswamy and Sundararaj (1957) to develop the technique. The dose given to the fish was 1½ homoplastic pituitary from the male donors, and 1½ pituitary from the females. Further works in this line have been done and it has been found that an extract from the carp pituitary gland, in 150-200 mg/kg body weight, bring about profuse spawning in 8-10 hours after injection. Thus *Clarias batrachus* and *Saccobranchus fossilis* can be easily cultured in stagnant or shallow waters in rural areas. Now they can be used for intensive fish culture in urban areas also due to availability of fish seeds, favourable response to artificial feeding provisions and fast growth.

It has been experimentally observed that in the polyculture practice with singhi (*S. fossilis*), magur (*C. batrachus*) and koi or climbing perch (*Anabas testudineus*) as combination of fishes, a production of 1,200 kg/ha within 7 months, without the use of supplementary feed and fertilizers, can be obtained on stocking young ones at the rate of 25,000/ha.

In Bihar it has been observed that when stocking of singhi and koi (climbing perch) was done in combination, the production obtained was 524 kg/ha after 8 months of culture period. This yield was at hand without the use of supplementary feed and application of fertilizer.

Singhi (*Saccobranchus fossilis*). This is another example of scaleless catfish inhabiting swamps, ponds and rivers (Fig. 18.6). The young fry of this fish feed on insects, ostracods, worms, algae, etc. Fingerlings, being carnivorous, feed on animal organisms co-living in water and show cannibalism also. Ramaswamy and Sundararaj (1957) got success in practising hypophysation with pituitary of the same species. But later it was found that injection of carp pituitary gland (a dose of 150-200 mg/kg body weight) brought profuse spawning after 8-10 hours of injection.

Like *C. batrachus,* the specimens of *S. fossilis* can be cultured in swamps, shallow water areas of rural area and in ponds of rural or urban areas as observed by several scientists including Bhatt (1968), Thakur (1978), Dutta *et al.* (1979). They can be used for monoculture practices as well as polyculture with the combination of *Anabas* spp.

Like *C. batrachus, S. fossilis* has high market value, the sale value per kg weight being higher to major carps. In several states including West Bengal, Assam, Bihar and Orissa, the availability of feed at ease, fast growth rate and, nevertheless, limited cost involved towards management and feed have brightened the scope of monoculture or biculture practices of these two catfishes.

Climbing Perch

This fish is represented by only one species, *Anabas testudineus* (Fig. 18.7). It is in great demand because of high nutritive value, and is widely distributed throughout South-East Asian countries including India. This fish attains maturity at 6 months, attaining about 9″ size, breeds in confined waters throughout the year, reaches a maximum length of 23 cm within a short span of time. The presence of pair of labyrinthiform organ (accessary respiratory organ) enables the fish to breath through accessary respiratory organ, besides normal gill-breathing and can remain alive out of water for a considerable long time (Fig. 18.8).

The fish is predatory in habit but not piscivorous like murrel fishes. Fry and fingerlings feed on phytoplankton and zooplankton, where as advanced fries, fingerlings and adults feed on crustaceans, molluscs, insect larvae, soft parts of aquatic plants (weeds) and dead organisms including dead fish.

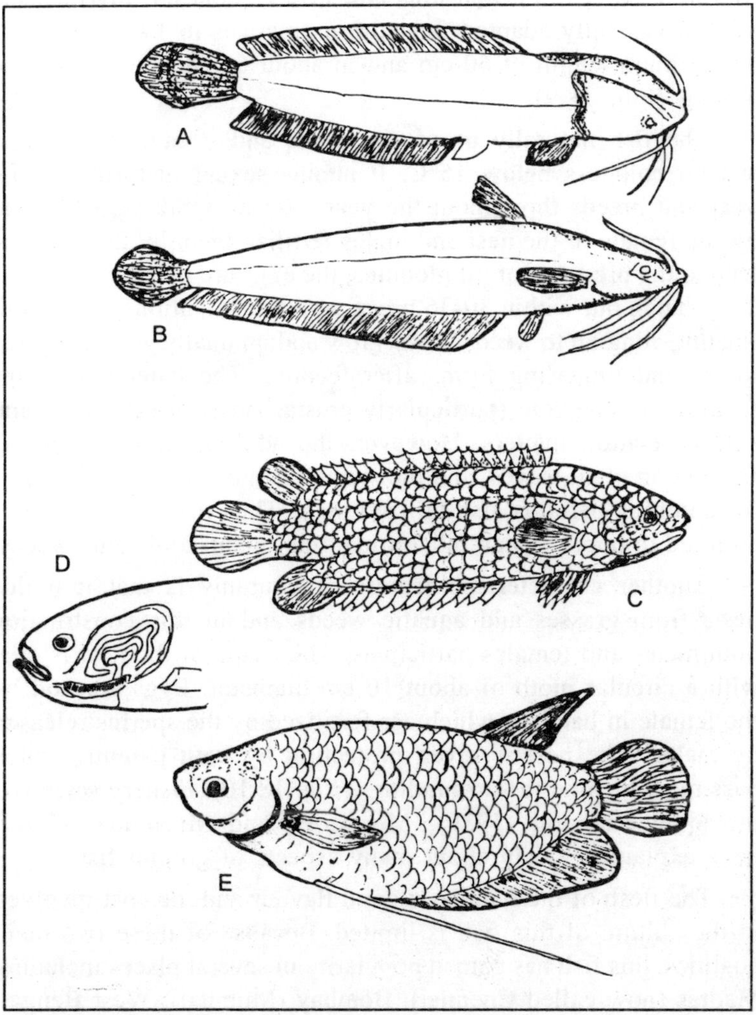

Fig. 18.5 to 18.9. Outline figures of some catfishes, *e.g.* magur, i.e., *Clarias batrachus* (A), singhi, i.e., *saccobranchus fossilis* (B), a perch fish, *Anabas testudineus* (C) and a gorami, *Osphronemus goramy* (E). In Fig. 18.8, i.e., of *Anabas* (C) labyrinth organ is exposed by sharp operation (D).

Giant Gourami (*Osphronemus goramy*). It has gained popularity and earned high demand throughout South-East Asian countries because of quality as a table fish. It was first brought to Tamil Nadu from Mauritius and Java (Indonesia) in 1865, but was successfully adapted for culture practices in 1916. It attains a maximum length of 60 cm and in about 6 years it weighs 4 to 4.5 kg (Fig. 18.9).

The fish naturally breeds in ponds and streams but cannot sustain coldness below 15°C. It attains sexual maturity in 2nd year and breeds throughout the year. 500 to 3,000 eggs are laid by the female in the nest and males fertilize them by their sperm. Due to the presence of oil globules, the eggs are pelagic in nature. They hatch out within 30-36 hours, and the hatchlings, soon after floating cling on to weeds. They grow and gradually get transferred to fry and fingerling forms after feeding. The latter two forms feed on zooplankton (particularly crustaceans), insect larvae and soft vegetation matters. However, the adult form is herbivore, feeding on plant leaves, and the supplementary feed as mentioned for climbing perch is also utilized by it. Thus fish is successfully cultured in ponds which inhabit marginal plants and some weeds.

Another characteristic feature of gouramy is that it builds nests from grasses and aquatic weeds and in this construction both males and females participate. The shape of the nest is oval with a circular moth of about 10 cm diameter. Eggs are laid by the female in batches, which are fertilized by the sperms released by males. The nest is given protection by both parents, which pass most of time in surroundings (Fig. 18.10). Fishery scientists including Kulkarni (1939), Mookerjee and Mazumdar (1946) have explained nicely about many aspects of gorami fish.

The flesh of the fish has a good flavour and the cost involved in the culture of this fish is limited. Because of these two main qualities, this fish has gained popularity in several places including Madras (now called Chennai), Bombay (Mumbai), West Bengal, Orissa and some other states. Out of these places, the fish has very high popularity in Tamil Nadu.

REFERENCES
(For Chapter 15, 16, 17, 18)

Alikunhi, K.H. 1953. Notes on the bionomics, breeding and growth of the murrel, *Ophicephalus striatus* Bloch. Proc. Ind. Acad Sci 38B (1) : 10-20.

——. 1956. Fish culture technique in India. Progr. Fish Developm, India : 63-73.

——. 1965. Observations on growth, maturity and feeding of induced-bred, pond reared silver carp, *Hypophthalmichthys molitrix* and grass carp, *Ctenopharyngodon idellus* in India during July 1962 to August 1963. Bull. Cent. Inst. Fish Educ. Bombay 2 : 19 p.

Armugam, T.M. 1966. Murrel Culture as a commercial fishery in Madras State. Madras J. Fish 2 : 54-61.

Banerjee, S.R. 1974. Hypophysation and life history of *Channa punctatus* (Bloch) J. Inland Fish. Soc. Ind. 6 : 62-73.

—— and Prasad. 1974. Observations on reproduction and survival of *Anabas testudineus* (Bloch) in Bihar region. *Ibid.* 6 : 6-17.

Bhanot, K.K. 1973. Observations on the spawning of *Hilsa ilisha* (Hamilton) in the Hooghly estuary J. Inland Fish Soc. India, 2 : 50-54.

Bhatt, V.S. 1968. Studies on the biology of some freshwater fishes. Part VII. *Heteropneustes fossilis* (Bloch). Indian J. Fish, 15 (1&2) : 99-115.

Bhimachar, B.S., David, A., and Muniappa, B. 1944. Observations on the acclimatization, nesting habits and early development of *Osphronemus gorami (Lacepdi)* Proc. Ind. Acad. Sci. 20 : 88-101.

Chacko, P.I. and Kurian, G.K. 1947. Culture of murrel fish, *Ophicephalus marulius* (Ham.) in irrigation wells. J. Bombay not Hist Soc. 47 : 392-93.

Chakrabarty, R.D. and Murty, D.S. 1972. Life history of Indian major carps, *Cirrhinus mrigala* (Ham.) *Catla Catla* (Ham.) and *Labeo rohita* (Ham.) J. Inland Fish Soc. Ind. 4 : 132-61.

——, Rao, N.S.S. and Sen, P.R. 1979. Culture of fish in ponds with fertilization In : Symposium on Inland Aquaculture, February 12-14, 1979 (Abs.) pp. 5-6, Barrackpore.

Devraj, K.V., Verghese, T.J. and Satnarayan Rao, G.P. 1972. Induced breeding of the fresh water catfish, *clarias batrachus* (Linn.) by using pituitary glands from marine catfish. Curr. Sci. 41 (25) : 868-70.

Dutta, A.K., Dasgupta, K.K., Pal, R.N. and Tripathi, S.D. 1979. Magur, *Clarias batrachus* (Linn.) Culture in rural ponds as a source of additional income. In Symposium on Inland Aqua-culture (Abstract) February 12-14, 1979 : 43 p CIFRI, Barrackpore.

Ganapati, S.V. and Chacko, P.I. 1950. Suggestions for stocking fish ponds in Madras. Madras agric. J., 37 : 1-5.

Ghorai, A.K., Mondal, K.M., Sengupta, A., Chakraborty, D., Randhir, M., Das, P. and Jhingran, V.G. 1979. Attainments of composite fish culture demonstration centres in West Bengal during 1976-77. Bull. cent. Inland Fish. Res. Inst., Barrackpore, 30 (mimeo) : 12 p.

Hora, S.L. 1938. A preliminary note on the spawning grounds and bionomics of the so called Indian Shad, *Hilsa ilisha* (Hamilton) in the Ganges. Rec. Indian Mus. 40 : 147-58.

——. 1941. Life history and wandering of hilsa in Bengal waters. J. Asiatic Soc. Beng. (Sci.) 6 : 93-112.

—— and Pillay, T.V.R. 1962. Handbook on fish Culture in the Indo-Pacific region. FAO Fish, Biol. Tech. Pap. (14) : 203 p.

Hosaini V.H. and Rahimullah M. 1946. Early developmental stages of *Ophicephalus punctatus* Bl. Proc. Ind. Sci. Congr. 33(3) : 126.

Imam, A.E., Roushady, H.M. and Philisteen P. 1970a. Feeding of catfish, *Clarias lazera* in experimental ponds. Bull. Insti. Oceanography in fish, Cairo 11 : 207-22.

Jhingran, V.G. and Sehgal, K.K. 1968. Coldwater Fisheries of India. pp. 239. Inland Fisheries of India, Barrackpore, India.

Jones, S. and Menon, P.M.G. 1951. Observation on the Life history of the Indian shad, *Hilsa ilisha*.

——. 1959. Fishing methods of the Indian shad (*Hilsa ilisha*) Hamilton in the Indian Region Part 1. J. Bombay nat. Hist. Soc. 54(3) : 714-25.

Karamchandani, S.J. 1961. On the location of spawning grounds in the freshwater regions of Narbada river. Curr. Sci. 30 (10) : 373-75.

Khan, H.A. 1972. Induced breeding of air-breathing fishes. Indian Fing. 22(4) : 44-45.

Kulkarni, C.V. 1939. Breeding habits and early stages of the gourami (*Osphronemus gouramy* Lacepede J. Bombay nat. Hist. Soc. 44 (2).

——. 1950. Breeding habits, eggs and early life history of the Indian shad *Hilsa ilisha* (Hamilton) in the Narbada river. Proc. nat. Inst. Sci. Ind. 16(3) : 169-76.

Lakshmanan, M.A.V., Sukumaran, K.K., Murty, D.S., Chakraborty, D.P. and Phillipose, M.T. 1971. Preliminary observations on intensive fish farming in fresh water ponds by the composite culture of Indian and exotic species. J. Inland Fish Soc. India 2 : 1-21.

Mac Crimmon, H.R. 1971. World distribution of rainbodat trout (*Salmo gaidneri*). Fish Res. Bd. Canada 28 : 663-704.

Malhotra, J.C., Mathur, P.K., Mehrotra, S.N. and Kamal, M.Y. 1973. Successful rearing of *Hilsa ilisha* (Hamilton) in free water for years. Proc. nat. Acad. Sci. India (B), 43 (3) : 207-10.

Mookerjee, H.K. and Mazumdar, S.R. 1946a. On the life history and of *Labeo Calbasu* (Ham.). J. Dept. Sci. Calcutta Univ. 2 (1) : 1-22.

—— and ——. 1946b. On the life history, breeding and rearing of *Anabas testudineus* (Bloch), *Ibid*. 2 (1) : 101-40.

Moser-Jefferson, F. 1899. The Salmon and Salmon Fisheries of Alaska. Bull. U.S. Comm. 18; 1898 : 178 pp.

Nilsson, Nils-Arvid, 1955. Studies on the feeding habit of trout and character in North Sweden Lakes. Rep Inst. Freshw. Res. Droltning, 36 : 163-225.

——. 1957. On the feeding habits of trout in a stream of North Sweden. Rep. Inst. Freshw. Res. Droltning, 38 : 154-66.

Parmeswaran, S. and Murugesan, V.K. 1975. Supplementary feeding and nursery rearing of murrel fry. Proc. nat. Acad. Sci. India 45B(2) : 133-42.

—— and —— 1976. Observations on the hypophysation of murrels (*Ophiceptialidae*). Hydrobiologia 50 (1) : 81-87.

Phillips, J.S. 1929. A report on the food of trout.

Pillay, S.R. Rao, K.V. and Mathur, P.K. 1962. Preliminary reports on the tagging of the *Hilsa ilisha* (Hamilton), Proc. Indo-Pacific Fish Coun. 10(2) : 28-36.

——. 1958. Biology of the hilsa, *Hilsa ilisha* (Hamilton) of the river Hooghly. Indian J. Fish 5 : 201-57.

Prowse, G.A. 1968. The latin square fertilizer trials. trop. Trop. Fish Cult. Res. Inst. Wkg. Pop. (1) : 28 p.

Raj, B.S. 1917. On the habits of Hilsa (*Clupea ilisha*) and their artificial propagation in the Coleroon. J. Asiat. Soc. Bengal, N.S. 13 : 184.

Sehgal, K.L. 1971 Report on the survey of coldwater fisheries in the South Indian States of Kerala, Tamil Nadu and Mysore (MS).

——, Kumar, Kuldip and Sundar, Shyam 1984. Food preferences of brown trout, *Salmo trutta* fario (Linnaeus) in five streams of Kashmir during 1971-74. Indian J. Anim. Sci. 54 (7) : 675-82.

Sen, P.R. and Chakraborty, R.D. 1979. Rearing of major carp fry to fingerlings in freshwater ponds. In : Symposium on Inland Agriculture (Abstract), February. Sidthimunka, A., Sanglert, J. 12-14 1979 : 18 p. CIFRI, Barrackpore and Pawapootanon, 1966. The culture of catfish (*Clarias* spp) in Thailand. FAO Fish Rep. (5) Vol. 44 : 196-204.

Singh, S.B., Sukumaran, K.K., Chakrabarti, P.C. and Bagchi, M.M. 1972. Observations on Composite culture of exotic carps. J. Inland Fish Soc. India 4 : 38-50.

Sinha, V.R.P., Gupta, M.V. Banerjee, M.K. and Dhirendra Kumar. 1973. Composite Fish Culture at Kalyani, West Bengal. J. Inland Fish. Soc. India 5 : 201-07.

Sundararaj, B.I. and Goswamy, S.V. 1969. Techniques of hypophysation of catfishes FAO/UNDP. Regional Seminar on induced breeding of Cultivated fishes. FRI/IBCF/, 15 : 22 p.

Tapiador, D.D., Henderson, H.F., Delmendo, M.N. and Tsutsui, H. 1977. Freshwater fisheries and acquaculture in China — A report of the FAO Fisheries (Aquaculture) Mission to China, April 21 to May 12, 1976. FAO Fish Tech Pap. 168 : 84.

Thakur, N.K. 1978. On the food of the air breathing catfish, *Clarias batrachus* Linn. Occurring in wild waters. Int. Reveges, Hydrobiol. 63 (3) : 421–31.

Tripathi, S.D. and Sharif, R.K. 1974. An artificial feed for air breathing fishes. Labdev. Journal of Science and Technology, India (B), 12 (4) : 151-52.

19

Cage Culture

Fish culture in cages has been practised since long past in certain parts of Asia. This culture practise is more relevant for those countries where suitable water supply and land for normal fish culture are not sufficient. For these reasons, fish culture in cages has become more popular in Japan and this culture is intensively practised for culturing both freshwater and saltwater species. The fishes commonly used for cage culture in Japan are yellow-tail (*Seriola quinqueradiata)* and sea bream (*Chrysophrys major*). It is estimated that the production of fish from floating cages alone amounts to over 90% of the total yield produced from the culture in shallow areas. Culture of carp (*Cyprinus carpio*) and other varieties of freshwater fish is also practised in cages on a large scale in Japan. The main reason of success in cage culture is the broken nature of the coastline which renders plentiful areas to have calm waters wherein marine fishes lobster and crabs are cultured in cages (Rounseefell, 1985). In Cambodia also, the cage culture has been practised from long time past.

In India, practice of cage culture is prevalent in West Bengal and Orissa, but the cage construction is mostly as yet in primitive form, *i.e.*, made up of split bamboo mats, bamboo splints etc. Realizing the advantages of cage culture, the Central Inland Fisheries Research Institute, Barrackpore sanctioned a co-ordinated project and, under this project, the cage culture is in operation in Assam, Bihar and Karnataka States.

Practice of cage culture is gaining popularity for many reasons. Some important reasons out of these are as follows :

(1) It requires less investment than on maintenance in traditional fish ponds for culture of fishes.

(2) Its installation is easy and, if need arises, it can be safely removed from one place to another.

(3) The introduction of fish in cages is easy and the fish farmer gets opportunity of feeding of fishes in cages, occasional inspection of fish regarding growth and health conditions and the treatment of diseases, if such a need is observed.

(4) Harvesting of fish in cages is not a problem at all, and the cost involved on operation of gears, manual labour cost, besides consumption of time, which usually comes up in the case of pond culture, are completely saved.

Construction of Cage

In India, the cage is in the shape of a square or rectangle or in any other shape as per choice. It is usually made of synthetic materials like polyester, polyamide, polyethylene, polyvinylidene chloride polyvinyl alcohol, etc. which has a flat bottom and an opening on the water surface and it has the support of long bamboo poles as displayed in the Fig. 19.1.

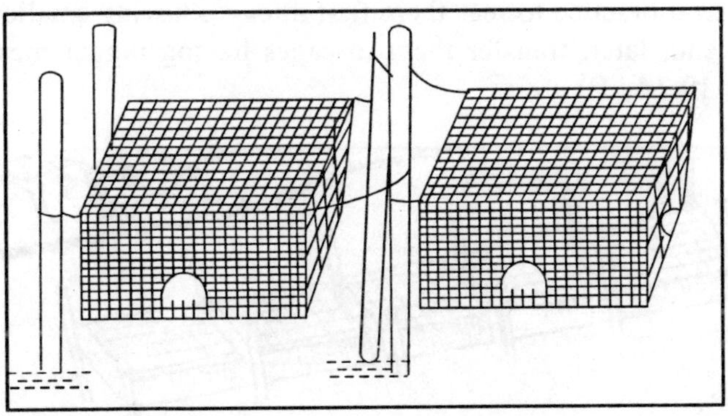

Fig. 19.1. A pair of cages made by split bamboo, popular in rural areas of Eastern India.

Now in commercial fish farms, bamboo poles have been replaced by steel frames with the result that the cages have better look, are stronger and much more durable. Japan is nowadays a leading manufacturer of synthetic net for constructing cages. During operation, the cages are kept afloat by a number of steel drums which serve as floats. However, steel drums are being gradually replaced by synthetic boats so as to further reduce the cost on cage construction, but retain the same efficiency. For preventing the fish from escape by jumping out, usually a wing net is provided which is kept attached all around the upper rim of the cage, as shown in the Fig. 19.1.

The surface area of the cage is varied, but 8-12 square metre is the standard size in which cages are constructed. The depth of the cage depends upon type of fish to be stocked. For the culture of salt water fishes in shallow areas, a depth of 3 to 5 metres is preferred, but for the carp culture in freshwater, two metres depth is an ideal requirement. Size of the mesh of the net which is generally 1½ inches or slightly bigger. Smaller mesh size is not recommendable on account of hindrance on water circulation, resulting in retarded growth of the fish.

Stocking in Cages

As fish cages have bigger mesh size (1½ inch or slightly more) only fingerlings or more grown up fishes are used for stocking in cages. If the collected seeds are in smaller size, it is a general practice to rear them first in cages having smaller mesh size, and, later, transfer them to cages having bigger meshwork (Fig. 19.2A, B).

Fig. 19.2A. Rows of cages for culture of air-breathing fishes.

Rate of stocking in cages is variable, being related to size growth and behaviour of fish to be caged. Thus in Japan, the standard stocking rate for stocking carp in cages is about 50 fish/m² surface area of the cage, that is, it amounts to 500,000 fish per hectare. In U.S.A. authentic recommendation for stocking the channel catfish, *Ictalurus punctatus* is 300 fishes

Fig. 19.2B. Rows of mat cages made of split bamboo for culture of air-breathing fishes.

in six inches size per cubic yard of water. In Cambodia, about 7,000 to 8,000 young pangasius fish, each in 80-150 gm weight, are introduced in a cage 7-10 m long, 2.5-4.5 m wide and 1.5-2 m deep. Consensus of opinion, based on yields, is that larger nets and longer periods of cage operation with sufficient feed lead to higher production from the cage culture (Fig. 19.3).

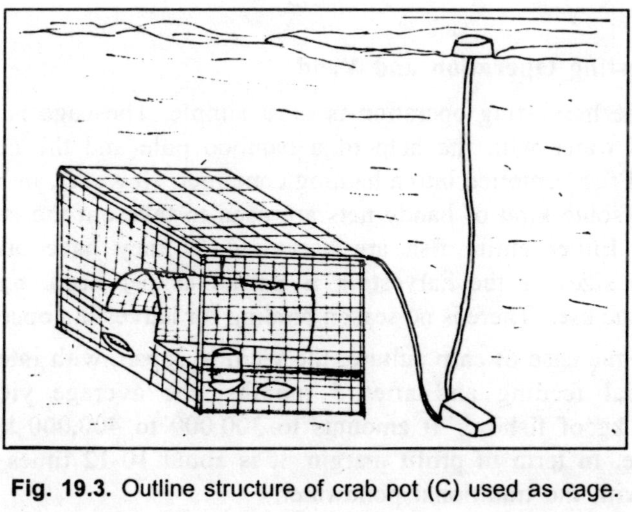

Fig. 19.3. Outline structure of crab pot (C) used as cage.

Feeding Scenery

Artificial feeds in the form of pellets are most commonly used in the cage culture practice. The pellet feed generally contains raw materials of animal origin (about 60%) and plant matters (about 40%). The animal matters generally consist of white fish meal along with meat, liver meal, bone-meal, blood meal and skin and milk, whereas the matters of plant origin are plant protein, rice bran, soyabean meal, etc. If the fishes stocked in cages are carnivorous, *e.g.* yellow-tail or sea bream, the chopped pieces of fishes of least market value, such as trash fish, horse mackerel (*Trachurus japonicus*), anchovy (*Engraulis Japonica*) sand lance (*Arctoscopus Japonicus*) are commonly used as feed (Nose and Toyama, 1966).

Feeding frequency in cages varies from fish to fish. In the case of carp culture, feeding is done 5-6 times a day but in the culture of yellow-tail, two feeds per day suffice the purpose. Sometimes provision of the light lures food organisms in cages, *i.e.*, attracts planktons in the cage during night. This device serves well for providing additional food to the enclosed fishes.

To minimise the cost on artificial feeding, cheaply available items such as rice bran, rice husk, pulse husk, mustard oil cake, trash fish, kitchen waste, slaughter house waste are being used and the results obtained are satisfactory in respect to involvement cost.

Harvesting Operation and Yield

The harvesting operation is most simple. The cage is lifted out of water with the help of a bamboo pole and the content (caged fish) emptied into a hauling container. However, in bigger cages, some kind of handy nets are used to take out the content (fish). Either entire fish are harvested (if they have attained proper size) or the harvesting is partial as per need, *e.g.* for domestic use. There is no season binding for harvesting operation.

In the case of carp culture, the yield in Japan, with intensive artificial feeding and after 5 months, the average yield is 30-40 kg of fish/m^2. It amounts to 300,000 to 400,000 kg per hectare. In term of profit margin, it is about 10-12 times more than with the traditional pondwater.

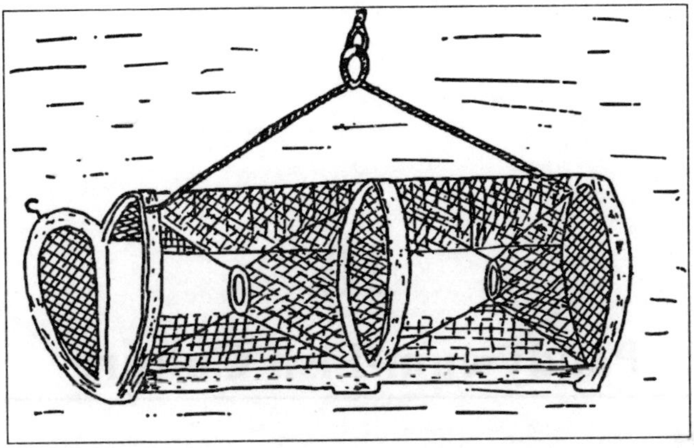

Fig. 19.4. A view of half round eel pot (EP) for entrapping eel type fishes and enabling their growth therein.

Concluding Remarks

The results obtained from the practice of culture in cages is encouraging and there is much scope for expanding the cage culture programmes on governmental, institutional and private agencies levels. The cage culture has opened a new alternative vista for easy exploitation of natural waters, both freshwater and saltwater. Management and investment of capital are minimum.

It is envisaged that this production-oriented and employment generating practice with low cost investment will receive more attention on international level. This device will provide additional source of fish protein from inland waters and there will be opening of additional avenue for the betterment of socio-economic condition of the rural mass.

20

Paddy-cum-Fish Culture

The culture of fish in paddy fields is an old practice in the rural areas of South-Eastern Asian countries. It can provide cheap and valuable protein food to the farmers and, thus, give an additional source of income so that economy of rural areas may get boosted up.

Paddy-cum-fish culture is very popular in Indonesia, Philippines and Madagascar (Africa) and in some other countries of South-east Asia and Africa. Nepal has introduced paddy-cum-fish culture in certain areas where common carp is cultured and this has now spread to larger areas with success. In India, this practice is more common in West Bengal and Kerala, where brackish water prawns and air-breathing fishes like murrels (preferably *Channa striatus*) are cultured. Iyengar (1953) recorded yield of 115 kg of fish per hectare after 122 days of culture in the paddy fields. Not only this, the yield of paddy increased by 7 to 13% after harvesting. Such encouraging results are available also from (i) the aqua-culture at Dhauli on the rearing of common carp fry along with China 1039 varieties (a hybrid) of paddy, and (ii) a paddy field of 0.16 ha at the Central Research Institute, Cuttack (Orissa).

In the state of West Bengal and Kerala paddy fields are generally flooded with fresh water. Some strains of paddy are resistant to brackish water which reach the paddy fields through earthen channels. In such brackish water fields both paddy and culture of euryhaline fishes like mullets, tilapia and brackish water prawns are raised successfully.

Method of Practising Paddy-cum-Fish Culture

Method. Experiments conducted on fish culture in paddy fields in West Bengal have shown excellent results with *Labeo rohita, Catla catla* and *Cirrhinus mrigala* fishes. These fishes exhibited more rapid growth in paddy field than in ponds. Another remarkable observation by scientists is that culture practices with *Channa striatus* in the paddy fields can yield average production of 112 kg/ha of fish and side by side, 7-13% increase in the yield of paddy may be obtained (Fig. 20.1).

Fig. 20.1. A scenery of paddy-cum-fish culture. PF = Paddy

The scientific procedure is that farmers fertilize the fields with organic manures before sowing or planting paddy seedling in the inundated field. Some of the nutrients dissolve in the water and spread all-around with the effect that production of plankton gets a boost up. Incidentally, the planktonic biota of the paddy field is mostly like those of shallow pond. Few additional suggestions for successful culture are given here.

1. As there is chance of drainage of water from the paddy field during rains through loose embankment it is desirable that strong bundh should be raised alaround and regular check up for any breakage due to cattle movement or otherwise should be made and repairing obtained (Fig. 20.2).

2. Earthen channels (by digging earth) and small pond-like (paddy-free) spaces should be constructed in the middle part of the paddy field so that shelter from sun's heat and fish eating birds or small fish eating mammals may be available to fishes. It

is desirable that width of the channel should be 50 cm and depth 39 cm. If the aim is to rear tilapia in the paddy field, the channel should have width of 1.2 to 1.8 m and depth of 60-90 cm.

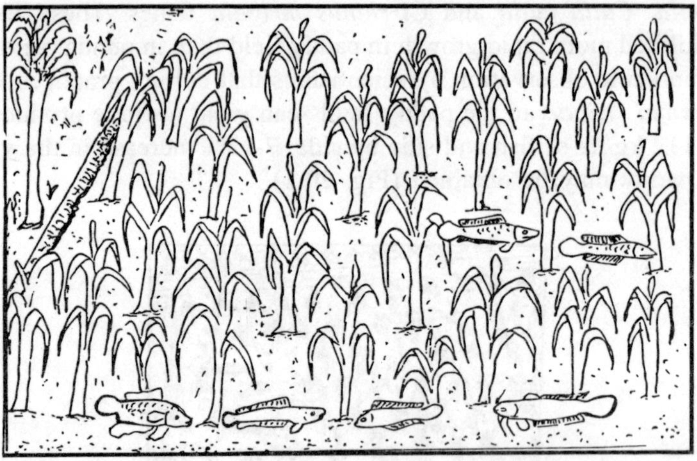

Fig. 20.2. Magnified view of a paddy field containing air-breathing fishes along with paddy plant.

3. Chance of fish for culture in the paddy field is generally based mainly on these considerations :

(a) This fish should be able to thrive in shallow waters.

(b) The fish should be able to withstand even fairly high level of turbidity of waters.

(c) This fish can sustain relatively high temperature and low oxygen content.

(d) This fish can grow to economic (*i.e.* marketable) size in a few months.

As common carp (*Cyprinus carpio*) satisfies all the requirement, it is preferred over other fishes for paddy-cum-fish culture. It is recommended that fries of 3-5 cm size should be introduced at the rate of 200 per hectare. If provision of additional food is made, the stocking rate can go up to 500 per ha. After a gap of 40-60 days, the carp yield is 30-50 kg per hectare, and after 9 months, the fish grows to 700-2950 gm (average 1.84 kg wt). Giving consideration to income from paddy yields as well, this yield of fish is a worthy bonus to farmers.

Fishes which can be cultured in the paddy fields with the prospect of profits are *Labeo rohita, L. fimbriatus, L. calbasu, Catla catla, Cirrhinus mrigala, Wallago attu, Clarias batrachus, Anabas testudineus, Channa striatus, Tilapia mossambica* etc.

Success in the experimental endeavour by Devraj and Natrajan (1973) in a paddy field in Tanjore district on stocking fingerlings of *Cyprinus carpio, Catla catla, Labeo rohita* and *L. calbasu* (stocking of about 5,000 fingerling/ha and production of 240 kg/ha after 9 months) have led to popularity of undertaking paddy-cum-fish culture in various areas of Kcrala. The growth of the specimen of *C. carpio* was spectacular, being of highest order. In 9 months, average gain in weight was 1840 kg and a survival of 49.8%. A suitable combination of fingerling of *C. carpio* (200), *C. catla* (1500), *L. rohita,* (1,000) and *L. calbasu* (500) at the rate of 5,000 fingerlings/ha gives a better yield. A specific suggestion is that murrels which are piscivorous/carnivorous should be kept out of the field by keeping guard at the inlets and outlets by proper setting of wire mesh-work.

The natural food in the form of minute crustaceans and insects are always availed. If manure is added (50-100% in addition to normal application of manure to have desirable extent of paddy growth), the availability of natural food increases (Chacko and Ganapati, 1952, Alikunhi, 1960; Devraj and Natrajan, 1973). The use of compost and night soil as manure are also helpful to increase natural food production, and resultantly bring faster growth of the fish. It is a remarkable fact that fish culture in paddy field, enhances growth of the paddy, as fish perform tillage, reduce insect population, destroy unwanted weeds that interfere in the growth of paddy plants. All these lead to the increase of paddy production.

Constrains

A matter of caution is that now high yielding varieties of paddy are cultivated, but these require use of insecticides, pesticides and weedicides of various kinds to control pests and weeds. Unfortunately these chemicals are very harmful for fish culture as they take life of various organisms including life of fries and fingerlings and natural food organisms of various kinds (Shafi and Akhtar, 1982; Akhtar, 1983; Sadhu and Shafi, 1993). Under such conditions, the fish culture practices with paddy cultivation is not possible. A consensus of opinion is that before undertaking any plan of fish culture in the field, expert advice should be obtained, giving consideration to soil and water conditions, the variety of paddy to be transplanted, depth of water in the field and presence and kinds of weeds present in the field. Condition of border wall, provision of outlet and inlet at the borderline should also be ascertained. Nevertheless, choice of fish kind to be cultured is also to be decided, basing on local condition of the paddy field, particularly in consideration to depth of water.

The fish under selection should be able to thrive in shallow water and sustain relatively higher temperature and turbidity. As already mentioned before, certain carps, murrels, catfishes (magur) and tilapia are preferably selected for culture in paddy fields in Indian conditions.

21

Integrated Fish Farming

Man's association with pet animals or cattle is an old habit existing since time immemorial. Archaeological records in different places like Khajuraho (in M.P.), caves of Ajanta, Ellora and Elephanta (in Maharashtra) exhibit this association. In rural areas man's living with a pond or ponds in nearby place and location of cattle farm, piggery, poultry unit, duck unit can be seen even now. Even banana plantation on the raised bundh of fish pond is done with a scope of profitable return. Co-living of man with these establishments is out of his own comforts of easy supervision, utilization of every inch of his land in possession and, at the same time, opportunity of availing fish, chicken, duck or cattle, and milk at his will. The income from sale of these animals besides marketing of harvested fish adds to the financial relief.

However, the traditional practice of co-living of village persons with fish ponds, cattle, chicken, ducks, goats, sheep, pig for certain facilities have been taken up by fishery scientists as an important issue and after getting success in the pursuit of experimentation scientific methods have been explored towards farming of fish culture dairy, poultry, horticulture etc. at a place.

In different parts of the world an integrated fish farming technology has emerged for getting maximum utilization of available land areas. In India also this practice has made progress, and the main concept for this undertaking is recycling of organic wastes to have more fish yields from fish ponds due to this manuring provision and to contribute in a big way towards state rural development programmes (Fig. 21.1). Two instances of integrated fish farming are as follows :

Fig. 21.1. Fish culture in pond along with piggery on its embankment.

1. The co-ordinated projects on composite fish culture, prepared and implemented by the Indian Veterinary Research Institute, Izatnagar (Uttar Pradesh) was recently completed with great success. It was concluded that the carp attained excellent growth even without adding any supplementary feed (cereal bran or oil cakes) and fertilizers (inorganic chemicals or organic fertilizers). In fact, the ponds receiving cow shed wastes on regular basis led to increase in the fertility of the fish pond. It was reported that the six major carps, when introduced into the fish pond with the stocking density of 7,200 fingerling/ha, the average growth in 10 months was as follows :

Name of the Fish	Weight gained	Name of the fish	Weight gained
Catla	2.96 kg	Silver carp	1.75 kg
Rohu	1.45 kg	Grass carp	2.30 kg
Mrigal	0.9 kg	Common carp	0.97 kg

Out of these, the case of grass carp was different as it was fed with aquatic vegetation. Other fishes were not given any sort of supplementary feed.

2. Operationed Research Unit on composite fish culture at Krishnanagar (West Bengal) recently made experimentation on fish culture in ponds, fed with pig manure. The fish pond was stocked with carp fingerlings and pig manure was added at regular intervals for 10-11 months.

The experimental pond was stocked at a density of 8,500 fingerlings/ha. The weight of the fingerlings at the time of introduction and weight attained by respective major carps were as follows :

Namc of the fish	Averagc weight of fingerling (in gm)	Weight attained after 10-11 months (in kg)
Catla	48	1.3
Rohu	28	1.0
Mrigal	23	0.9
Silver carp	9	2
Grass carp	5	2
Common carp	—	0.6

The grass carp was given grass of berseem and hybrid Napier variety to suit its habitual feeding demands. No supplementary feed was introduced into the pond, nor any inorganic fertilizer was added. The overall result obtained, as reported by Sharma *et al.* (1979) was promising in consideration to the finance incurred.

In China, the prevalent practice is to use only organic manure in the fish pond. With this consideration, the fish farmers are hired to undertake intensive animal husbandry near ponds. An expert advice has been given by Sinha, 1981; Sinha and Ramachandran (1985) in connection with integrated fish farming. The fish pond receiving organic manures from cattle or other animals or birds should be desilted preferably once in 3 to 4 years, depending upon the extent of accumulation of organic matter at the bottom. It is a remarkable fact that the silt so removed is very rich in organic matters and other plant nutrients. As such, it should be used as rich manure in agricultural use. In fact, the practice of using silt removed from fish ponds is already prevalent in certain areas, particularly in eastern part of India. The experiences for utility of silt from sewage fed ponds have been employed in the integrated fish farming as well.

Fig. 21.2. Fish culture in pond with cattle-keeping *e.g.* cow on its embankment.

It is a recognised fact that zooplankton constitutes the principal food of carp spawn. When a water body is manured, there is production of a rich crop of both microplanktons and macroplanktons. The macroplanktons feed on the microplanktons and grow well, forming the food of fishes, and this leads to increase in the overall productivity of water body. Alikunhi (1963) suggests that the cow-dung is one of the best easily available organic manure which results in thick swarms of zooplankton. Many other scientists including K.V. Huilgol and M.R. Thosar have confirmed that there is high growth of micro-zooplankton on treatment of water with cow-dung and this leads to rise in the population of macro-zooplankton. Axon (1963) has experimentally shown that the survival of spawn is directly related to the quality and quantity of zooplankton present in the water, and concern of plankton production with some hydrobiological factors have been shown by Moitra and Bhattacharya (1965).

In the integrated fish farming programme the dung or night soil of cattle *i.e.*, of cow, buffalo, sheep, goat, pig, chicken, duck have potentiality to enrich the productivity of fish pond by high

Fig. 21.3. Fish culture in pond along with poultry farm (PF)/duck farm (DF) on its embankment.

production of zooplanktons which are used by carp spawn and fry as food for their growth (Figs. 21.2, 21.3). It was, therefore, advisable to give due weightage to integrated fish farming programmes and open scope of socio-economic upliftment of the common people. Nevertheless, integrated programme of freshwater prawn culture with pig farming represents an efficient way of earning money out of smaller investment. The details of fresh water prawn culture are given in the next chapter, entitled 'Fish Culture in Brackish Water Ponds.'

Fig. 21.4. Fish culture in pond along with horticulture on the nearby land.

REFERENCES
(For Chapters 19, 20 and 21)

Akhtar, M.N. 1983. Histophysiological studies of the effects of two agricultural chemicals on the Vital Organs and Blood of some paddy field Fishes of Ranchi. Ph.D. Thesis, Ranchi University, Ranchi.

Alikunhi, K.H. 1960. Rice field Fish culture outside Indonesia. In : Lectures presented at the Third International Inland Fisheries Training Centre, Bogor, October-December, 1955.

Axon, 1963. Paddy-cum-Fish Culture. Indian Fish Bull. 10(3) : 7.

Chacko, P.I. and Ganapati, S.V. 1952. Fish Culture in paddy fields. Ind. Com. J. 1; 173-75.

Devraj and Natrajan, V. 1973. Experimental paddy-cum-fish culture in Arupathiodae Tanjore district. J. Inland Fish. Soc. Vol. V : 46-49.

George, M.J., Mohamed, K.H. and Pillai, N.N. 1968. Observations on the paddy field prawn filtration of Kerala, India. FAO Fish. Rep. 57(2): 427-42.

Gopinath, K. 1956. Prawn culture in the rice-fields of Travancore-Cochin, India, Proc. Indo. Pacif. Fish Coun. 6 (18) : 419-25.

Hora, S.L. 1951. Fish culture in rice fields. Curr. Sci. 20(7) : 171-73.

—— and Pillay, T.V.R. 1962 Handbook on fish culture in the Indo-Pacific region. FAO Fish Biol. tech. Pap. (14) : 204 p.

Iyenger, H.D.R. 1953. Paddy-cum-Fish Culture. Pilot studies conducted at Hesserghatta and Viswewaraya Central Farms in Mysore State. Ind. J. Vet. Sci. and Animal husbandry 23 (4) : 289-97.

—— 1962. Further studies on paddy-cum-fish culture at Hesserghatta fish farm in Mysore State. J. Bombay nat. Hist. Soc. 59(1) : 301-05.

Kask, John L. Yashio, Hiyama. 1947. Japanese fishing gear. U.S. Fish and Wildl. Sene., Fish, Leaf 234, also Rept. 71, Nat. Resources Section, Supreme Commander for the Allied Powers : 107 pp.

Lerdprasert, M. 1964. The culture of Clarias in floating basket. Thai Fish Gag : 13(4).

Moitra, M.A., Pandey, Asha Kumari, Ghosh, T.K. and Datta Munshi, J.S. 1979. Spawning behaviour, past embryonic development and culture of *Anabas testudineus* (Bloch) In. Symposium on Inland Aquaculture (Abstracts), February 12-14, 1979 : 2 CIFRI, Barrackpore.

Pillay, T.V.R., Bose, B.B. 1957. Observations on the culture of brackish-water fishes in paddy fields in West Bengal (India). Proc. Indo-Pacif. Fish Coun. 7 (2 and 3) : 187-92.

Rounseefell, George A. 1985. Artificial propagation of Salmonoids as a management tool. In "Fishery Science," International Book and Periodical Supply Service, 24 B/5 Desbandhu Gupta Road, Karol Bagh, New Delhi-100005.

Sadhu, D.N. and Shafi, Md. 1993. Occurrence of histopathological lesions in the gill epithelia of *Anabas testudineus* exposed to Dimecron (a pesticide) treated media. 73rd Indian Sci. Congr. Assoc. 1986 : 98.

Shafi, Md. and Akhtar, Md. Naseem. Histochemical Studies on the effects of Folithion 50 on the hepatic tissues of a murrel fish, *Channa Striatus* (Bloch). Proc. 69th Indian Sci. Congr. Assoc. Pt. III. p. 82.

Sharma, B.K., Dilip Kumar, Das M.K., Das, S.R. and Chakraborty, D.P. 1979. Observations on Swine-dung recycling in composite fish culture. In : Symposium on Inland Aquaculture (Abstract), February 12-14, 1979 : 99 CIFRI, Barrackpore.

——, Dilip Kumar, Das, M.K., Das, S.R. and Neogi, S.K. 1979. Observations on fish-cum-duck rearing. *Ibid.* : 95.

Sinha, V.R.P. 1981. Integrated synergic approach to aquaculture. Resource Management and Optimization, New York, U.S.A. 1(4) : 331-41.

—— and Ramachandran. 1985. Freshwater Fish Culture. p. 1-80 Indian Council of Agricultural Research, New Delhi.

Tripathi, S.D. 1963. Fish as a byproduct of paddy cultivation. Indian Livestock 1(2) : 53-54.

Umali, Augustin F. 1950. Guide to the classification of fishing gear in the Philippines. U.S. Fish and Wildl. Serv. Res. Rept. 17 : 165 pp.

Verma, C.P. 1958. A note on Paddy-cum-fish Cultural experiments in Bihar during 1957. Indian Fish Bull., 5(4): 11.

Woynarovich, Elek. 1975. Elementary guide to fish culture in Nepal. FAO of the United Nations, Rome, 131 p.

22

Fish Culture in Brackish Water Ponds

Fish culture in brackish water has gained great popularity in recent decades. There is less cost involvement in the culture practices, and the products, particularly prawns and shrimps, have high marketing value mainly due to excellence in taste and high nutritive merits. In India, the culturable brackish waters are spread in 14.22 lakh hectare areas, along the coast line of West Bengal, Orissa, Andhra Pradesh, Tamil Nadu, Kerala, Karnataka, Goa, Maharashtra and Gujarat.

For culture practices in brackish water ponds, foremost requirement is the construction of pond at such a suitable site where good supply of the tidal waters can be regularly available. Preferably brackish water ponds are constructed near the sea, adjacent to the mouth of estuaries, which receive salt water at every high tide. George (1962) is of the opinion that a tidal range of 1.5 m is essential for a brackish water fish farm. Another requisite of a brackish water pond site is the clayey soil whose presence in the pond bottoms, raise scope of production of natural food, particularly planktons in sufficient quantities, as reported by Matida (1967) after investigational works. Clayey soil is suitable also for the embankments around the fish pond.

The simplest type of pond is in the form of a sheet of water, spread in an area of 0.4 to 2.4 hectare and having water in a depth of 0.3 to 1.3 metre. The pond is fed with water through wooden or bamboo sluice which controls the entry and exit of water. In Philippine, the size of a brackish water pond is usually large, being 10 hectare in area with provision of nurseries and transition ponds in the same farm. Generally 1 per cent of the pond area is reserved for nurseries, 9 per cent for transition ponds and the rest portion, that is 90% area is meant for stocking ponds. However, in Hong Kong, the brackish water pond has basically no difference from the fresh water pond. In such ponds, generally mullets are cultured along with carps.

Pond Construction

The lay out of the pond is basically the same as that of fresh water ponds. However, a special emphasis is given to build a strong embankment all around. The height of the embankment is kept in accordance with the height of the tidal water in the adjoining sea. The latter is, in most cases, 30 cm above maximum flood level. In case of larger brackish water ponds the slope of the pond bottom is so adjusted during construction that the water may drain out easily. However, in the deepest portion of the farm, the sluice gates are constructed to control water flow. In west and central Java, simple wooden or bamboo gates with side walls and shutter boards are fitted in a frame and placed in the position of sluice gates. In India, however, wooden sluice boxes with wooden shutters are used. The shutters are such that they can be easily manipulated by hand as per need of occasion. Usually a 'V' or 'W' shaped bamboo grating is constructed behind the sluice gate in India. This provision helps in controlling the entry of undesirable fish and other animals.

Quality of Soil

The role of soil is of great significance in productivity of pond, even greater than in a fresh water pond. Nature of the land drained by the river determines the quality of the underlying soil of a brackish water pond. In Bengal, the clayey loam soil of deltoid region is very fertile. The incoming tides bring organic matters from the sea which get deposited on the bottom soil, and this practice consequently increases the fertility. In well established ponds, several kinds of plants species, such as *Closterium, Pediastrum, Azobactor, Nostoc* and *Anabaena* grow which have natural ability to fix atmospheric nitrogen in brackish water pond. Resultantly, nitrogen content of the soil is greatly enhanced by the nitrogen contributed by these plants.

Physico-chemical Features

The physico-chemical features of the brackish water pond is slightly different from that of the fresh water pond. These differences are mainly in respect to turbidity, temperature, pH, dissolved oxygen content and biotic organisms.

Turbidity. The turbidity level shows a wide range of variation. It is a common feature that once a pond is constructed, the silt is deposited quickly due to stagnant conditions created. This situation, *i.e.,* suspension of silt in water, increases the turbidity level. However, this situation is short-lived, as during a short time, there is admixture with fresh water and, consequently, flocculation occurs. Thus turbidity level gets reduced.

Water temperature. The water temperature of the brackish water pond does not fluctuate much. As the pond is shallow, the question of stratification does not arise. Usually a range of 14.4°C—35°C is found in the brackish water ponds of West Bengal.

pH. The water in brackish water pond is invariably alkaline due to containment of fresh water in bulk. There is little variation in pH. As the alkaline reserve is high, there is no chance of wide fluctuation of pH.

Dissolved oxygen. The temperature and salinity affect the oxygen content in a positive way and depletion of oxygen is of rare occurrence in brackish water ponds. The riverine like condition of water generally keeps the oxygen in saturated condition in the brackish ponds and planktonic organisms have a flourishing life in such a condition.

Water supply. There are two main sources of water supply in brackish water ponds (i) rain water, and (2) tidal water from the estuaries. During rainy season, there is accumulation of water which dilutes the contained water of the pond and decreases the salinity. However, during dry or summer months, the evaporation as well as seepage of water cause loss of water. These occurrences increase the salt content. The incoming tidal water from the sea increases the salinity, and with the return of tidal water, the same decreases. Thus, there is diurnal fluctuation of salinity. The diurnal (through tidal water) and seasonal changes (through rain water in the monsoon and drought condition during summer) in the salinity level greatly influence the growth of aquatic biota and mostly the surviving species are only euryhaline groups of organisms.

Biotic communities. In brackish water pond several genera and species of *Myxophyceae, Diatomeceae, Chlorophyceae* grow luxuriantly and algal population dominates. Several types of bacteria also grow on the bottom mud. The fish culturists of Philippine have recognised two types of algal population in brackish water ponds : (1) lab lab, in which *myxophyceae* algae predominates, and (2) lamut, in which *chlorophyceae* have predominance (Lin Kwei, 1991; Tsai, 1991).

Among the fauna, crustaceae such as ostracods, cladocerans, copepods are most common and mysids and decapod larvae are also found in large numbers. Though planktonic organisms are in plenty, there is less utilization of these as the animal, feeding on planktons, particularly zooplankton, are very few. Among macro organisms, fishes are in abundance. Species of prawns and shrimps, such as Penaeus, Metapenaeus and Leander are found in plenty. Gobiid fishes are also found on the muddy banks. Among predatory carnivorous fish, Lates, Threadfins and Megalops are most common in Indian brackish water ponds.

Fertility of the Pond

The fertility of the pond can be increased by methods as practised for fresh water ponds, *e.g.* by (i) regular draining and drying the pond bottom as these devices can increase the fertility. (ii) Secondly, the green manuring can increase the population of the plankton, *Myxophyceae.* In Java and Philippines, the fish culturists often transplant algal growth and add some artificial fertilizer (as nutrient) to enhance algal growth. In Hongkong, the pond is drained first, then manures in the form of the rice bran, peanut cake, pig manure and night soil are applied. Water is then allowed to enter the pond, and the pond water is left as such for a fortnight before stocking with fingerlings. Lin (1991) and Tsai (1991) have given valuable informations about the state of aquaculture in Philippines.

Pre-stocking Requisites

The most widely cultured species in brackish water ponds are milk fish (*Chanos*), mullets, and prawn. If the pond is having a low salinity, fishes like common carp, grass carp, silver carp, big head mud carps and tilapia are also stocked. The stocking rate is like this in different countries :

In Taiwan	: 8,000 to 9,000 fingerlings per hectare
In Philippines	: 1,000 to 2,500
In Indonesia	: 1500. Stocking is done twice or thrice a year.
In India	: There is no regularity for stocking, but by admitting tidal water several species of prawns, and fish, such as milkfish, mullets, pearl spot get entry in South Indian brackish water ponds, whereas in West Bengal, mullets, Lates and prawns have dominance over other fauna.

For the maintenance of water depth in a required level, the embankment must be strong and channels as sluice gates must be in excellent working condition. In ponds having tidal flow of water, there is no difficulty in renewing and controlling the flow of water in the ponds. As the salinity may increase to a dangerous level in some brackish water ponds, it becomes imperative to flush the ponds with fresh waters.

Artificial feeding. Rice bran is considered as most suitable artificial food for the milkfish and mullets. Small quantities of rice bran are fed to growing milkfishes and mullets. However, algae form the most important food items for these fishes. In Hongkong there is regular system of supplementary feeding in the following order :

Rice bran	: 2000 kg/ha
Soyabean cake	: 500 kg/ha
Peanut	: 36 kg/ha

This combination of supplementary feed is applied to young ones of mullets and carps and satisfactory growth rate is achieved.

Stocking of Pond

Proper selection of species is very important consideration for stocking in brackish water ponds. For practising improved farming in these ponds and providing proper management, the following requisites need consideration :

(i) Availing of seeds in sufficient quantities.

(ii) Proper hardiness of the cultural species to sustain fluctuations in the environmental conditions, such as temperature and salinity.

(iii) Ability for rapid growth.

(iv) Power of compatibility with other species in consideration to consumption of food.

(v) Ability to utilize natural food (*e.g.* planktons) of the pond and also accept supplementary feed.

(vi) High conversion efficiency for conversion of consumed food into flesh.

(vii) High survival rate.

(viii) High market demand.

As following species of prawn and fishes fulfil the above criteria, they are preferred for brackish water pond culture.

1. Prawns (Fig. 22.1) like *Penaeus monodon, P. indicus, P. semiculatus, Metapenaeus monoceros, M. dobsonii, M. affinis, M. brevicornis*.
2. Herbivorous fishes (Fig. 22.2) : *Hilsa ilisha, Nematolosa nasus, Pellona* spp., *Setipinna* spp., *Perches* (*Lates calcifer, Etroplus suratensis, E. maculatus, Epinephlus* spp., *Lutjanus* spp., threadfins (Eleutheronema tetradactylum, *Polynemus* spp.), milkfish (*e.g. Chanos chanos*).
3. Carnivorous fishes (Fig. 22.3) : *Arius* spp., *Pangassius panagassius* and *Mystus* spp., *Osteogeniosus militaris* etc.

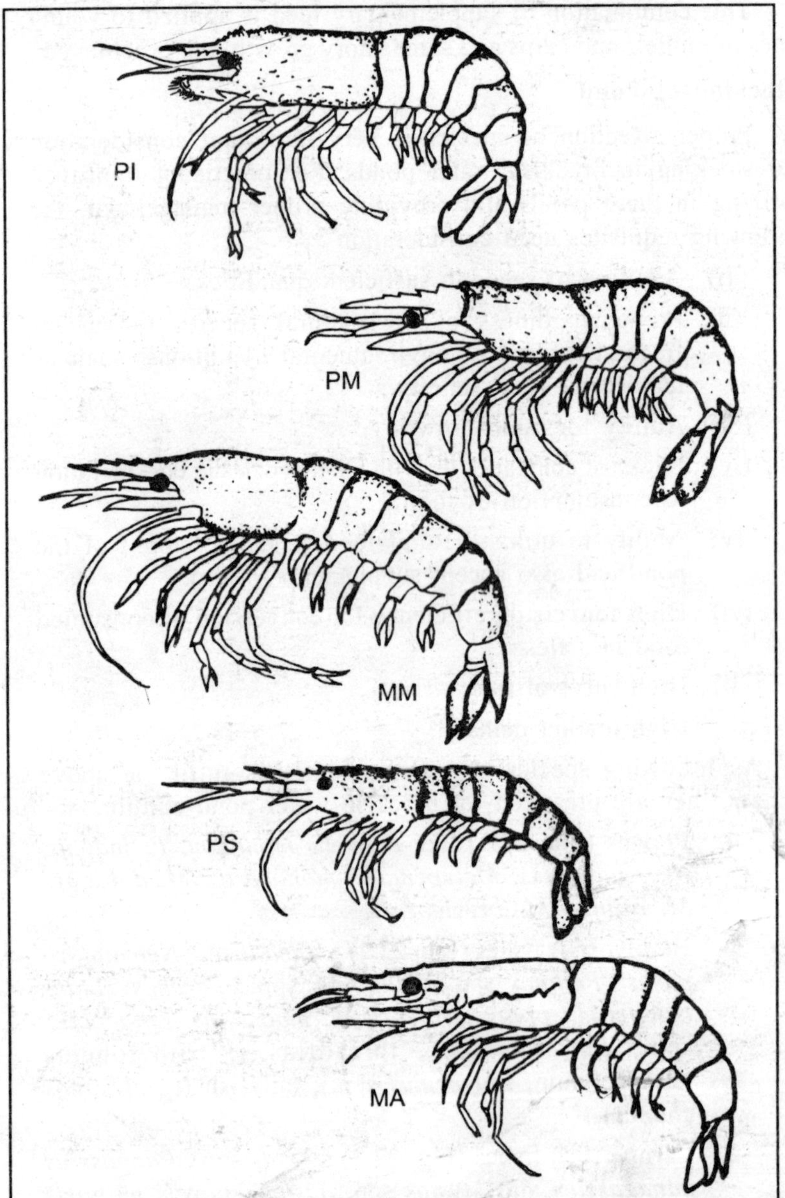

Fig. 22.1. Commonly found Indian fresh water prawns; *Palaemon Indicus* (PI), *P. Monodon* (PM), *Metapeaeus monoceros* (MM), *Parapenaeopsis stylifera* (PS) *and M. affinis* (MA).

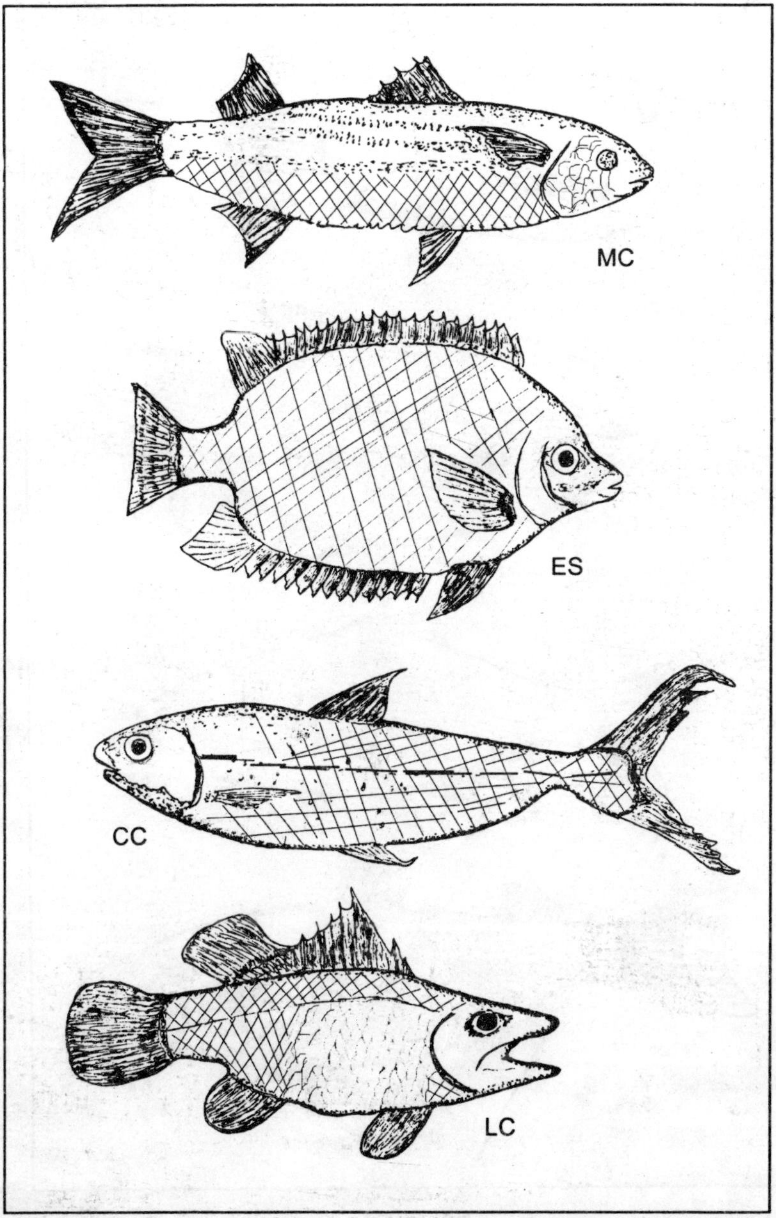

Fig. 22.2. Commonly found herbivorous fishes : *Mugil cephalus* (MC), *Etroplus saratensis* (ES), *Chanos chanos* (CC) and *Lates calcifer* (LC).

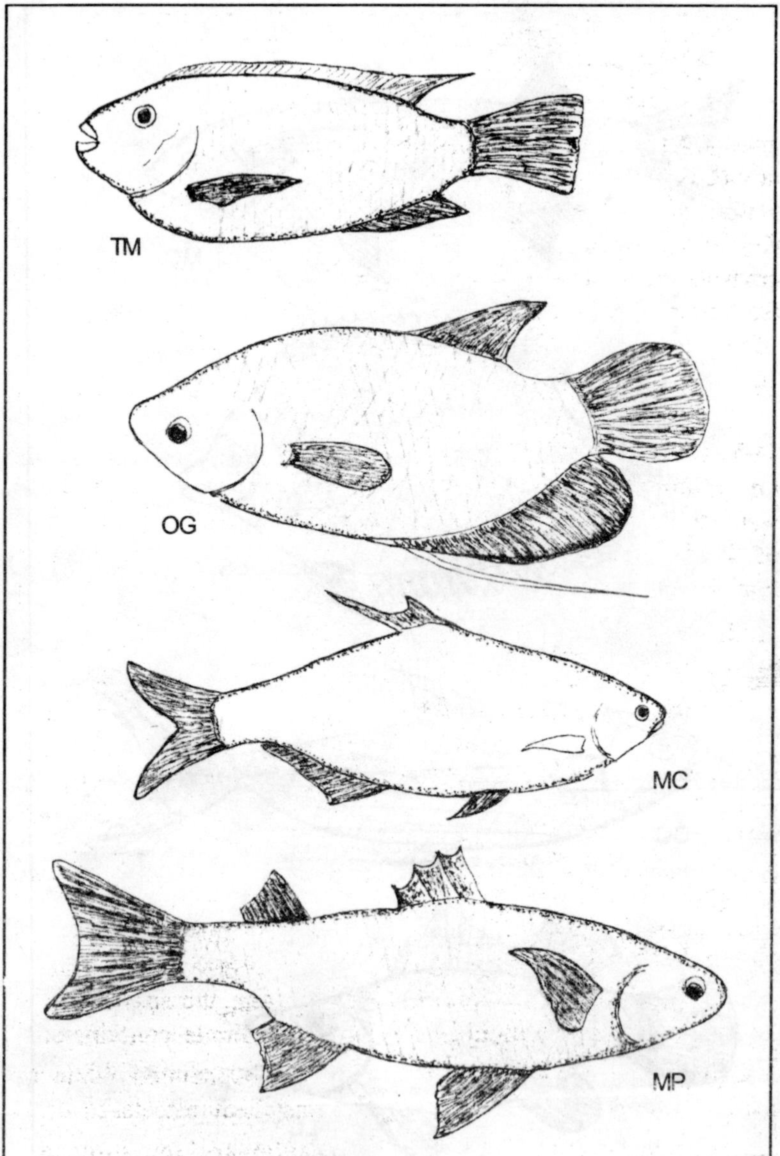

Fig. 22.3. Commonly found Indian carnivorous fishes : *Tilapia mossambica* (TM), *Osphronemus goramy* (OG), *Megalops cyprinoides* (MC) and *Mugil parsia* (MP).

Natural Feeds for Brackish Water Fishes

Basic characters of some natural feeds of the brackish water pond fishes are as follows :

1. *Natural Food of Animal Origin* : These are mostly the following : Cyclops, *Moina, Chaoborus* larvae, Chironomus larvae, P. rotifers (Wheel animalcules), *Keratella, Brachionus, Filinia, Daphnia, Sida, Nauplius, Ceriodaphnia, Eurycerus, Chydorus, Macrothorix, Polyphemus, Copepoda, Diaptomus.*

   Natural food of Plant origin (Diatoms) : Thc main diatoms are the following; *Tubellaria Diatoms, Meridion, Fragilaria, Synedra, Eucoconeus, Nitzschia, Pleurosigma, Amphipleura, Anomomoeoneis Navicula, Rhizoselenea, Cascinodiscus, Melosira, Cyclotilla, Achnanthes, Chaetoceros, Amphora, Caloneis, Diploneis.*

Estuaries (river mouths) and backwaters are ideal places for selected fish culture practices. Fishes suitable for brackish water culture include milkfish (*e.g. Chanos chanos*), pearl spot (*Eutroplus suratensis*), Bhetki (Lates calcifer), tarpon (*Megalops cyrprinoids*) and exotic tilapia (*Tilapia Mossombica*) and gouramy (*Osphronemus goramy*) (Fig. 22.4). The culture of milkfish (*Chanos chanos*) and the mullets (*mugil* spp.) is more popular because of easy cultivation, rapid growth, less mortality due to tolerance of wide ranges of salinity, and temperature, and giving high yields. They are used as monoculture practices or as polycultures with *Tilapia* or *Puntius* spp., in several South-east Asian countries. The yield is between 200-300 kg/ha/year. In India, *Mugil cephalus* constitutes 70-75% of the stock of catches. It has been found to grow to 1-1.5 kg in a year.

Success in fish culture practices in brackish water is related to ecological conditions. The most important aspect is the factor of varying salinity. It is an interesting fact that inflowing river water and rainwater tend to dilute salinity but the entry of tidal water and evaporation increases salinity. Certain areas may become partially dry in some season *e.g.* summer. There is fluctuation of temperature in the length and breadth of water areas. It may be high or low as per depth of water and the general climatic conditions of the surroundings. Requirements of organic and mineral fertilisers, supplementary feeding and other measures have already been given before. Thus the culture of fish in brackish water is gaining popularity and new informations are pouring in for a brighter scope of this culture.

23

Prawn Culture in Fresh Water

With increasing demand for prawns in the international market, prawn culture in brackish water ponds has gained great attention in last two decades. In India there are about 1.422 million hectares of brackish water areas which are suitable for the production of prawn and fishes. Besides, there are 4200 hectares of kharlands along the coast of Karnataka which provide great scope for taking up culture of prawn and of brackish water fish.

Out of several species of prawn, the most popular prawn in India is the fresh water giant prawn, *Macrobrachium rosenbergii* which is commonly known as 'chengri' in West Bengal and Orissa (Fig. 23.1). This species is traditionally cultured as monoculture practice, but also for polyculture along with major carps. It is commonly distributed in the tropical and subtropical countries of Indo-Pacific region including their sea waters. In India this species occurs in natural form in the estuaries and fresh water zones of river mouths and back-waters having 0 to 20 ppt. salinity range. Another species, *M. malcolmsonii*, is the next popular prawn gaining full attention for culture practices.

This species (*i.e., M. rosenbergii*) was introduced from Malaysia to Hawaii (USA), Japan, Thailand and several other countries of South-east Asia. However, the farming of this prawn has become much popular on Hawaii and Thailand. In India this prawn has sizable natural population in West Bengal, Kerala and Orissa. Also Maharashtra, Gujarat, Andhra Pradesh has good population of this giant prawn (Raman, 1964, 1967; Subramanyan, 1975).

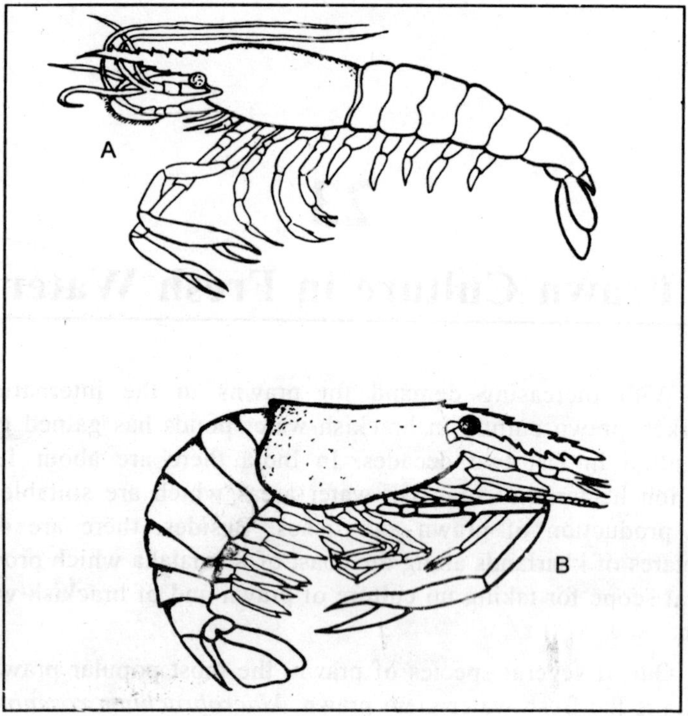

Fig. 23.1A,B. An outline view of the adult form of prawn *Palaemon rosenbergii*.

Among the commercial catches from estuaries and fresh water bodies the prawn is conspicuously present and it carries high commercial value. It attains an average length of 150-250 mm, but maximum size achieved is 320 mm. The sexes are separate and they attain maturity in about 4 to 5 months under proper conditions. The males are bigger in size and females produce 40,000 to 1,50,000 eggs. On average, 70,000 eggs are produced, each of which measures 0.6 to 0.7 mm in length. The eggs are bright red in colour, but the hatchlings are greyish. After 11 days, the colour turns slaty. The growth of prawn is very fast. The larvae pass through 11 larval stages during a period of 28 to 37 days, then they become post-larvae which settle to the bottom and become benthic in habit, feeding upon rotifera, atremia, nauplii of crustaceaus, tubifex, worms, fish eggs and egg custard.

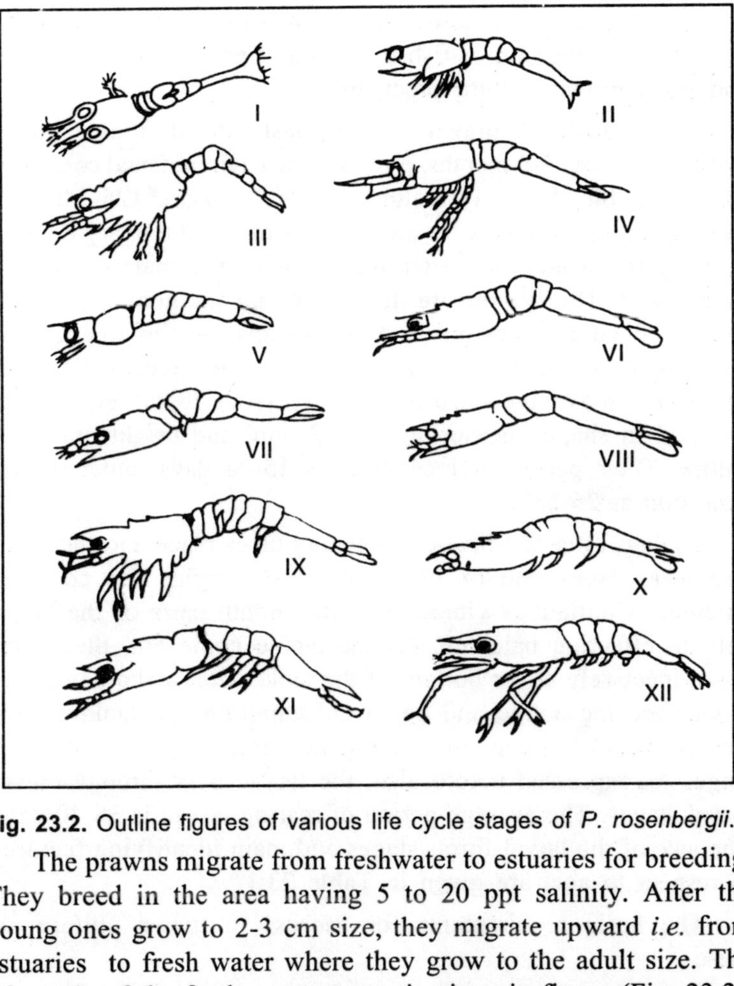

Fig. 23.2. Outline figures of various life cycle stages of *P. rosenbergii.*

The prawns migrate from freshwater to estuaries for breeding. They breed in the area having 5 to 20 ppt salinity. After the young ones grow to 2-3 cm size, they migrate upward *i.e.* from estuaries to fresh water where they grow to the adult size. The life cycle of the fresh water prawn is given in figure (Fig. 23.2), which reflects its phases in both fresh water and brackish water and the same has been observed experimentally by several biologists including Meeran and Sebastian (1973) and Subramanyan (1975). The popularity of the fresh water prawn has increased recently in consideration to many reasons including (a) high growth rate, (b) omnivorous feeding habits (as bottom feeder), accepting varieties of feeds, (c) high nutritative value, rich flavour and taste (d) scope of handsome earning due to high demand in the market. For these reasons, the giant fresh water prawn has attracted the attention of fish farmers and private firm owners to undertake prawn culture practices.

The growth of prawn is very fast and it attains sexual maturity in about 4-6 months, provided the environmental conditions are favourable. Sexually mature females measure 125-140 mm, and within 36-40 days after attaining maturity, the ovary develop fully in freshwater just after mating with the males, which is short-timed. The sperms are deposited near the genital pore and later fertilization takes place. A female spawns fins to six times after maturity and the fertilized eggs are deposited in the brood chamber. An average sized female lays about 70,000 eggs which are oval in shape, measuring 0.6-0.7 mm, and bright orange in colour. Their period of incubation is 15-24 days under normal condition at 26-28°C.

20 days after fertilization, hatching takes place and the entire egg mass takes one or two consecutive nights for complete hatching. Continuous vibration of the mouth parts of the larvae help in effecting hatching. As the larvae come out, they settle down inactively at the bottom of the tank. After a short time, the larvae become active and begin consumption of plankton. The earliest form of larvae, after emergence from the egg, is the zoea stage. As expressed before also, the larvae pass through eleven larval stages. The average period of rearing larvae is 36-40 days. The size of the larval forms/stages and main identifying features, in respect to age, are given in Table 23.1.

The basic requirements for success in prawn culture are these :

(i) availability of prawn seeds (juvenile forms) in plenty,

(ii) existence of suitable temperature.

(iii) containment of suitable quality of water in the rearing tanks, and

(iv) provision of suitable type of feeds for conversion to body protein.

In prawn culture practice, 28°C temperature is considered as adequate for larval development. However, experimental results

Table 23.1

Stages	Age/days	Size/mm	Main feature
I	1	1.92	Eye sessile
II	2	1.99	Eye stalked
III	4	2.14	Appearance of uropod
IV	7	2.50	Dorsal teeth on rostrum; 2
V	10	2.80	Telson elongated but narrow
VI	14	3.75	Appearance of pleupod
VII	17	4.06	Biramous pleupods
VIII	20	4.88	Pleupods bearing setae
IX	24	6.07	Endopods of pleupods with intema.
X	28	7.05	On rostrum; 3 or 4 dorsal teeth.
XI	31	7.23	Rostrum with many teeth on half upper dorsal margin.
XII (Post-larvae)	36	7.69	Both upper and lower margins of rostrum provided with teeth.

have shown that the larvae of prawn can grow well in the temperature ranging from 26.5 to 31.5°C.

Method of Prawn Culture

Prawn juveniles (*i.e.*, post-larvae in 2 or 3 cm size) are procured from rivers with the help of scoop net by direct operation or by using traps made up of bunches of various bushes. The bushes are tied with the monofilament or coir ropes are fixed at water depth of 5-6 feet during high tide in the river generally for 3-10 days. After expiry of stipulated time, the bush trap is lifted up and a gentle shake is given to it after spreading a cloth in 5 × 3 feet size below. As a result to shake, the juveniles of prawn fall on the cloth from the components of bushes, where juveniles had taken shelter from the tidy river water (Fig. 23.3). The juveniles thus collected are transported to the former tanks generally in open plaster pools of 5 × 4 feet or 6 × 3 feet size.

Fig. 23.3. Outline view of the juvenile form of prawn, *P. rosenbergii.*

The stocking ponds is prepared by dewatering, deweeding and drying. If drying is not feasible, manuring is essential in the form of mahua oil cake application at the rate of 250 to 500 kg/ha with pond water. Two weeks after application of mahua oil cake or direct after drying the pond, the pond is filled with water upto a depth of 1 metre. Generally 250 kg/ha lime is added in the pond. However, if cow-dung and single superphosphate are requisitioned for application, they should be given at the rate of 10,000 kg and 250 kg per hectare. If mahua oil cake is applied the cattle-dung amount should be reduced by 50%.

After 5-6 days of manuring, seeds are stocked at the rate of 50,000 number per hectare. The density of stocking may be increased to 1,50,000 number per hectare, if intensive culture is desirable. In the latter case, the feeding rate with supplementary feed is enhanced substantially. The stocked larvae are fed with different types of feeds such as custard of egg, artemia, artemia nauplii, fish flesh, tubifex worms and nauplii. The adults prefer to feed on grain, insects, worms, molluscs, crustaceans, fish, boiled eggs, carrot and cabbage (in crushed forms).

A proper management of prawn culture gives average yield of 45-47 kilograms per hectare. This rate of production is considered as encouraging in consideration to high demand both inside the country and in the export markets. If the culture of prawn is taken up along with major carps in tanks, additional income to the fish farmers will be available. Usually after 4-5 months, harvesting is done with the help of cast net, seine net or directly by hand-picking.

24

Prawn Culture in Brackish Water

Prawn has high market value due to delicious taste and high nutritive value. Due to marketing value, there is unending demand for catching large quantities of prawn from the coastal waters. However, in some areas only small and undersized prawns are caught in large quantities. To meet the ever rising demand the prawn culture practices have been taken up in India on a large scale in the last decades and prawn production reached a figure of 2,95,622 tonnes in 1991. Out of these penaeid prawns were 1,55,340 tonnes and non-penaeid prawn were 1,40,282 tonnes. India has 14.22 lakh hectares of brackish water (1995 data, source state governments/UTs) which are suitable for prawn culture.

Influencing Factors for Prawn Culture

The specific factors stipulating prawn culture in India are mainly as follows :

(i) Prawns have rapid growth and they reach marketable size within a span of 3-4 months only.

(ii) By undertaking culture practices as per standard technology, a production rate of about 1,000-1,500 kgs/ha/year could be achieved. This would fetch high price value.

(iii) Culture practices of penaeid prawns can be undertaken in different types of ecosystems, such as salts pans, derelict waters, shallow canals in the coconut groves and profitable harvest can be obtained.

(iv) As the technique is simple, it can be taken up even by unskilled, *i.e.,* illiterate farmers who can have self-employment with a scope of substantial profit.

Desirable Environmental Requirements for Prawn Culture

Successful practice of prawn culture has the requirement of following environmental factors :

(a) The pond should have shallow soft muddy bottoms provided with enough organic matters, detritus and other food items. These food items are items of choice by several species of prawns such as *Penaeus duorarum, P. Japonicus* and *P. indicus.* Sandy bottom of brackish water ponds provide enough food.

(b) Proper water quality is a foremost criteria for a successful prawn culture. Water should be maintained as pure as far as possible, free from acute turbidity and any pollutants; otherwise prawn stock would be reduced due to occurrence of mortality. There should be free exchange of water with highly oxygenated sea water, generally above 3.5 ppm. This preventive measure reduces mortality and boosts up growth of prawns.

(c) Penaeid prawns can sustain wide range of salinity. The following minimum salinities should be maintained for better yields of prawn, as per expert advice.

Penaeus indicus	—	4 ppt
P. Semisulatus	—	19 ppt
Metapenaeus affinis	—	14 ppt
Parapenaeopsis stylifera	—	25 ppt

Out of all these, *P. indicus* is most choiced specimen as it can be reared in still lower salinities. However, *P. monodon* is another choiced prawn for culture practices as availability of its larvae (fry) is in abundance.

(d) For a proper prawn culture practice in a fresh water pond, temperature requirement is 28°-32°C. For keeping the temperature in this range it is essential that there should be a regular exchange between the tidal water and pond water.

(e) For getting fast growth and moulting of the prawns, availability of sufficient amount of calcium in the pond soil, and pond water is to be ensured. This would facilitate moulting process.

(f) Availability of seed at proper time is one of the foremost requirement for culture of prawn in a large scale. Young ones of marine prawns, such as *Penaeus monodon, P. indicus, P. marginensis* are found in abundance in tidal creeks and coastal shallow water during specified seasons. In the case of *P. monodon,* two periods of larval development were recorded in the Chilka lake by Ramakrishnaiah (1973) : during January-February and June to October, whereas Subramanyan and Rao (1968) observed January to April and August to November as rich incursion of this species in Pulicat lake. Prawn seeds are collected from there and sold to interested farmers. The collection of *P. monodon* seeds from fresh water sources is possible but bulk supply to cover needs of large scale prawn farming operations is a difficult task, as several days are taken for accumulation of required quality of fish seeds. Moreover, a considerable level of mortality occurs during storage of seeds.

The requirement of seeds is fulfilled through controlled breeding of prawns in hatcheries followed by mass rearing of larvae. Thus adequate supply of prawn seeds of uniform size and of good quality is possible by this method to cover needs of commercial prawn production by culture practices in brackish water ponds.

Collection of Breeding Prawns

The marine prawns mature and breed in the sea. It is generally found that they utilize the adjoining lagoons and estuaries as nursery grounds where they grow to juveniles and semi-adult forms. It is well-known that lagoons and estuaries form the feeding grounds for penaeid prawns where these enter the post-larval stage and form a fishery of immense importance.

Trawl hauls kept for short duration in running sea water on deck pick up gravid females in living condition with spermatopheres implanted in thelycum. In the trawl net, mother prawns are caught along with other prawns but the mother prawns are easily identified due to the presence of butterfly or diamond shaped ovary close to its head region (Fig. 24.1). As the mother prawn is delicate, it is picked up carefully and wrapped up individually in a piece of soft net material and its both ends are secured with rubber bands. In wide mouthed plastic bucket, containing sea water, mother prawns are placed and the latter can be easily transported to desired places. If the destination is at a distance, seemingly to consume several hours, it is advisable to replace the used sea water with fresh one, collected from the same fishing ground.

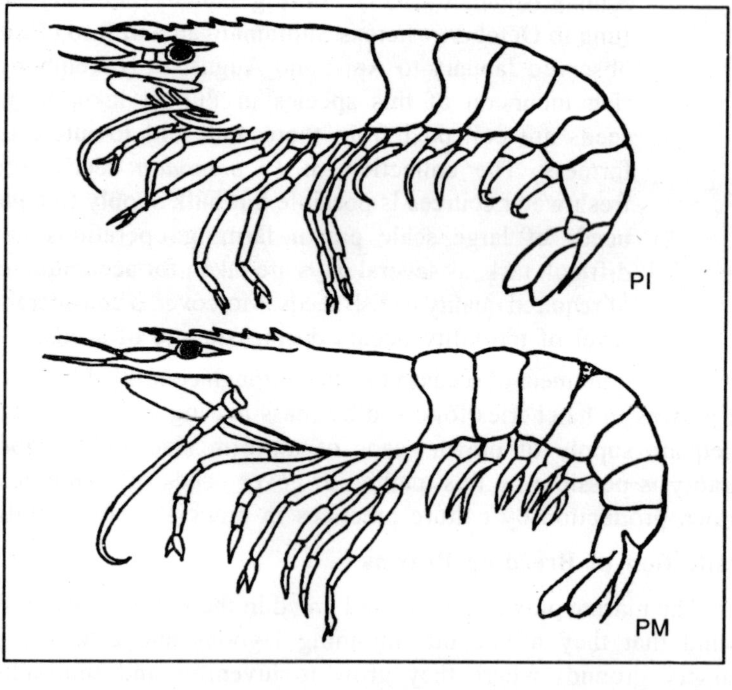

Fig. 24.1. Outline figures of the tiger prawns, *Penaeus monodon* (PM) and *P. indicus* (PI).

Other method of collection is like this. The mother prawns are placed in perforated PVC pipe with the help of soft nets and PVC clips. These are packed in polythene bags filled with air or oxygen (Fig. 24.2). The polythene bags should be immersed in a water tank so that the mother prawns can be saved from excess heat, generated during transit.

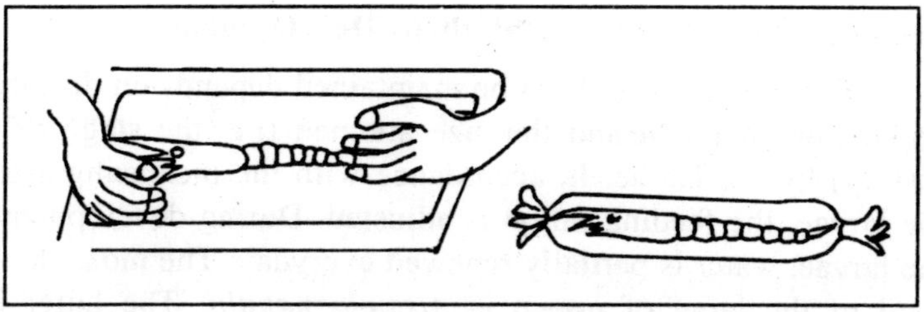

Fig. 24.2. Mother prawns packed in polythene bags.

The mother prawns after transportation are released in specially prepared spawning tanks containing clean fresh filtered sea water (having salinity range 28 to 34 ppt) and provided with artificial aeration at several points. If the mother prawns, *i.e.*, spawners are ripe, they spawn overnight and the eggs hatch out into nauplii after 12-14 hours.

Next step is to remove the spawners, increase the depth of water to 60-70 cm. After 36-42 hours, the nauplii undergo sixth moult and metamorphose to the larval form, called zoea. After another 3-4 days, the zoea undergoes, in normal water conditions, third moult and transforms into the mysis stage (Fig. 24.3A).

Fig. 24.3A. Larval development of a tiger prawn. *P. monodon* comprising Nauplius stage (N), Zoea stage (Z) and Mysis stage (M) in sea; Post-Larva stage and juvenile stage in river estuaries. MP=maturation of prawn.

Feeding of the Larvae and their Development

The feeding schedule to be maintained depends on density of population of larvae and the size attained (*i.e.* the stage of life history) by the larvae. In accordance with the increasing size of the larvae, the feeding ration is adjusted. During development of the larvae, water is partially renewed everyday. The most choiced food of the larva of prawn is *Atremia nauplii.* The latter is a shrimp, the living nauplii of which constitute the choiced main food of the prawn larvae *i.e.,* mysis. After 3-4 days, the mysis moults into the first post-larval stage. After eggs of prawn hatched into the nauplii, it takes about 9 to 10 days to reach the first post-larval stage. The latter is kept in the nursery tank for another four days. During this period, the aforesaid feeding schedule with *Atremia nauplii* is maintained. On the fifth day, the larva can be harvested and transferred to specially built nurseries for rearing.

Rearing in Nurseries

During rearing, proper feeding and maintenance of hygiene in the hatchery tank is necessary so that there can be proper growth and occurrence of less mortality. The ideal stocking ratio is 50 to 1000 per litre of water so as to get good yield of prawn.

The production of *diatoms* in the culture medium is increased by the application of inorganic fertilizers. These algal forms are consumed by developing larvae of the prawn. During later stages, *Atremia* eggs and nauplii, besides rotifers (*Brachionus* spp) and water fleas (*Moina* spp) are readily consumed by the larvae. As such, nutrients in proper quantity and kinds should be available in the media to influence their growth.

Atremia salina form excellent food for early development of prawn. It has been found that they have two main development stages : nauplii of 0.2 mm and metanauplius of 2.6 mm sizes. Nauplius requires 48 hours for hatching but 66 to 95 hours are required by metanauplius for hatching. After 10 days the larvae attaines length of 10 to 12 mm, and after 12 days, it attains a length of 12 to 15 mm which is the adult form size. Mature gravid *Atremia* carry cysts with them.

Induced Maturation and Spawning of Penaeid Prawn

Collection of mature prawns, that is, spawners from the sea during peak spawning seasons (March to June) is not difficult, but during other seasons, specially during the monsoon months, it is easy to obtain spawners.

The collected spawners are subjected to careful abalation and cauterisation of eye-stalks. The prawns receiving operation are placed in cement tanks or plastic pools containing sea water. They are given most suitable environmental condition and food provision is made to an adequate extent. All these favourable conditions lead to rapid gonadal development, *i.e.,* gonadal maturation. It has been observed that the spawner specimens of *P. monodon* attain full ovarian maturity within 25-30 days after operation. These induced matured specimens spawn large number of viable eggs. They grow and hatch to yield active nauplii. The larvae are reared in small enclosures like cement tanks or plastic pools and the method of rearing is the same as described before for spawners obtained from sea.

Prawn Seed Hatchery in India

There are two main prawn seed hatchery units in India, functioning well to produce prawn larvae.

1. Marine Product Export Development Authority (MPEDA) has set up a prawn fish hatchery unit at Gopalpur in Orissa with the technical assistance of M/s France Agriculture with a capacity to produce to an extent of 25 million prawn larvae (PL 20) per annum.

2. The second hatchery of MPEDA has been established at Mangamaripet near vizag in Andhra Pradesh with the technical assistance of M/s Aquatic Farms, Hawii to produce 40 million PL 20 prawn larvae in a year.

Culture of Tiger Prawn, Penaeus Monodon

Among penaeid prawns, *Penaeus monodon* has been most favoured prawn for prawn culture practices in India (Fig. 24.3B). It is least suited for brackish water culture. The favour to this prawn is because of its large adult size, fast growth rate, hardy nature, for its greater capacity of tolerance to varying salinity range and availability of its seeds due to natural abundance.

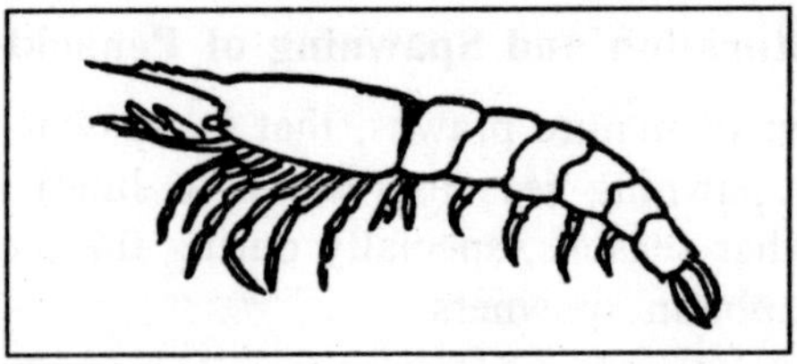

Fig. 24.3B. Juvenile form of *P. monodon* (a magnified view).

Seed Availability. The prawns breed in the sea, and their seeds move into estuary or brackish pond water along with tide, as post-larva of 10-14 mm size. The seeds of this specimen are found in plenty in the lower Sundarbans in the Hooghly-Matlah estuarine system, Chilka lake, Gopalpur backwaters, Mahanadi estuary, Gautami-Godavari estuary, Adyar estuary in Tamil Nadu. The seeds in post-larval form are available throughout great part of the year, with a peak during March-June. However, juveniles are available during June-September in the Sundarbans.

Seed collection. There are several methods of collection of seeds from natural sources. Use of shooting net, scoop netting and screening at gluice gate of embankment are three means of collection.

Shooting net. It is commonly used in Midnapore district of West Bengal. It has a net of 1/16″ mesh, with fine meshed open tail piece (Fig. 6.4). It is also used to collect major carp spawn in rivers. In new moon and full moon days during spring tides, maximum collection is obtained before the tidal water floods the intertidal zones. It is estimated that as many as 32,000 post-larvae of *P. monodon* can be collected with shooting net in 90 man hours.

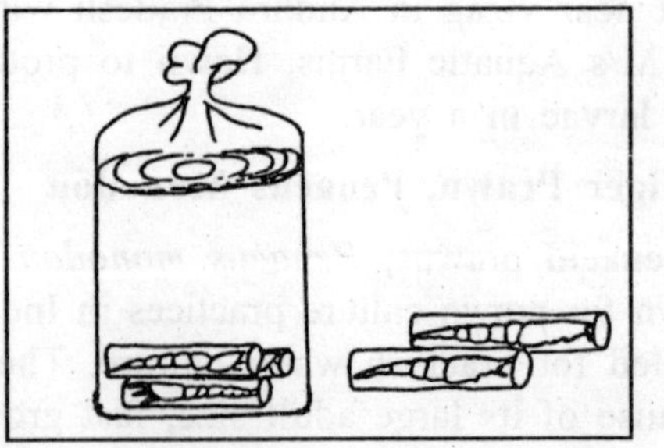

Fig. 24.4. A view of transportation of prawn in polythene bags.

Scooping net. When the tidal water ebbs, the post-larvae and fry settle in intertidal creeks and lagoons, and, then, they are easily collected with the help of scoop nets or hapa nets. The hapa nets are made of mosquito curtain cloth in the dimension of 2 m × 1 m × 0.5 m (Fig. 15.2). The collected seeds are processed carefully for segregation of seeds of *P. monodon*. An improvement in the collection tank has been brought by digging artificial pits of 4 m × 2 m × 0.4 m size in the intertidal beds near the farm. Bundles of weeds and twigs are kept in the pits. Such pits are easily strained with hapa nets after every high tide.

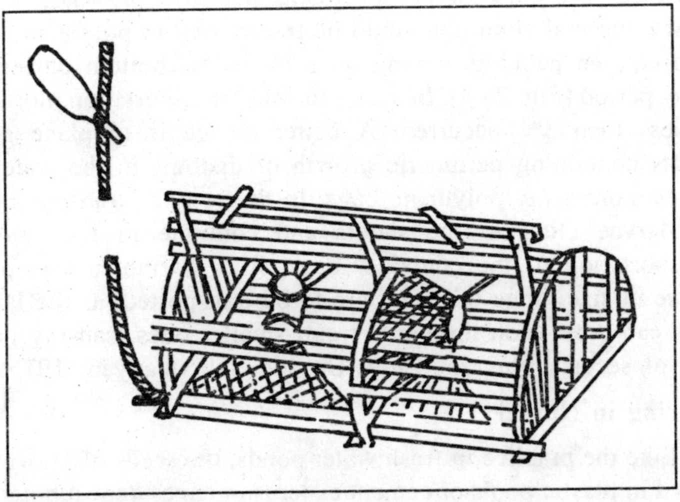

Fig. 24.5. A lobster cage used for capturing lobster and other crustaceans.

Screen at sluice gate. It has been observed that the tidal ingress through sluice gets of the brackish water pond naturally brings a good number of prawn seed into the feeder canal and subsequently to the ponds. By applying screening device at the sluice gate, the post-larvae and juveniles of *P. monodon* get entrapped and collected for rearing purpose.

The larval incursion of *P. indicus* and *P. monodon* occurs mainly at night in Pulicat lake and Godavari estuary, but in Chilka lake there seems to be no clear phenomenal indication of such incursion by these two penaeids (*P. indicus* and *P. monodon*). It has been observed that during winter spring, the larval incursions occurred mainly during night, and in summer and autumn during the day time.

Transportation of Seeds

From the stock of collected specimens of species of prawns, the seeds of penaeid prawn, *i.e.*, of *P. monodon,* are sorted out and placed in open containers for transportation to short distances. Early juveniles of 18-26 mm size at the temperature of 30°-32° C could easily be transported with 54 numbers of specimens per litre of water in safe condition for a duration of 3 hours. Likewise, 100 post-larvae specimens per litre of water, under similar thermal condition could be transported in polythene bags with oxygen packing, having no folds at the bottom during 18 hours period (Fig. 24.4). In such protected transportation mortality @ less than 3% occurred. A better device is to place grass shoots containing peripheric growth of diatoms in the water of the container; *i.e.* polythene bags. In this way of transportation, post-larvae cling to the shoots and continue to feed during transportation of the container bags. It is warranted that post-larvae in size range of 5 mm are to be transported, as the larger ones can prey upon the weaker and smaller ones, causing gross loss of seeds (Jhingran, 1969; De and Subramanyan, 1975).

Rearing in Nursery

Like the practice in fresh water ponds, the seeds of prawn are reared in plastic pools or cement cisterns or cages kept submerged in feeder canals.

Plastic pools of 400 litre capacity are filled with diluted estuarine water, maintaining the salinity at 10-12 ppt. The water is then exposed to sun with traces of urea till the water turns green with the growth of *Chlorella*. The post-larvae are introduced into this green water. The stock of post-larvae is maintained @ 10,000 per cu. metre water for 2 weeks. After this period, the density is reduced to 5,000 post-larvae per cu. metre and after 4 weeks, the density is reduced to 2,500 numbers per cu. metre. All these endeavours lead to the attainment of the post-larvae to 45-50 mm size after 2 months. Following these procedures, the survival is to an extent of 60-70% of the initial stock.

Cage Rearing

This is another device for rearing the early juveniles of the prawn. The cages are kept submerged in feeder canals and the juveniles are introduced into these from the plastic pools or cement enclosure or those directly collected from river. A design of the cage is shown in figure (Fig. 24.5). As the water from feeder canals directly mixes with the water in this case, water remains pollution free and survival rate is high. Addition of weeds and grass in the cage provides substratum to the juveniles and the natural growth of periphytic diatoms provides food to the developing larvae.

The stocking of the early juveniles is @ 1000/m^3. Maintenance of larvae in the cage for a period of one month leads to growth of the fry to 40-50 mm in total length, *i.e.*, attainment of, on average, 0.5 gm weight. The survival rate is as high as 95%.

Food and Feeding

P. monodon is an omnivorous prawn feeding on all sorts of animal and vegetable matters like plankton, bottom fauna like amphipods, isopods, nereid worms etc. Post-larvae and early juveniles enjoy consuming diatoms and other algal matters from periphytic growth on grass weeds.

Supplementary feeding. Post-larvae of *P. monodon* devour feeds, in powdered form, of different kinds, prepared under different formulations. However, the most accepted formulation is a mixture of dried and powdered offal goat, prawn in powdered condition, wheat flour, algal powder and yeast in the ratio of 20:2:40:15:5 by weight with traces of cobalt chloride (0.05 ppm) and terramycin (200 ppm). The conversion ratio was observed to be 2:1. It has been found that early juveniles attain a monthly growth 27 to 30 mm and 0.8 to 1.0 g by weight.

Rearing in Pond : Post-stocking Management

As mentioned earlier for nursery ponds, rearing pond is drained off and dried so as to cause eradication of unwanted organisms, particularly carnivorous fishes and predaceous aquatic insects. If it is not possible to dry out completely the pond, the water level is lowered to 10 to 15 cm and the trash fish and prawns are removed with the help of net. Repeated ranking of the bottom and draining of excess water helps in removal of abnoxious gas.

Organic manures in the form of cow-dung/poultry manure @ 6,000 kgs/2,000 kgs per hectare in a single instalment is added to the pond salt. Next course of action is to allow tidal water into the pond to stand at a depth of 20 to 30 cm for a fortnight. This endeavour helps for the growth of thick green algae and the bottom soil will develop algal mat containing mainly filamentons blue-green algae and associated organisms. The water level is increased gradually to reach full capacity.

Stocking of fry. Juveniles of *P. monodon* in 40-50 mm size are introduced in the stocking pond. This constancy in size range will reduce chance of cannibalism and will maintain uniform growth pattern. If weaker fries are there, they would have no chance of survival as they all fall prey to stronger individuals.

The recommended rate of stocking is 3000/ha, but in a larger pond/the rate may be increased to 50,000/ha. If we consider 40-50% as occurrence of mortality, this stocking density will yield production of prawn @ 300 to 350 kg/ha per crop (harvest).

Stock Pond Management

The principal requisites for the pond management as per expert advice are as follows :

1. Water quality should be like this :

Dissolved oxygen	:	8-9 ppm.
pH	:	around 8.2-8.4
Salinity	:	upto 10 ppt.

Any fall of salinity level below 10 ppt, *e.g.* due to heavy rains, *i.e.* accumulation of rain water, may be compensated by allowing the water to drain off and new tidal water is frequently allowed to flush the pond. Regular action of raking bottom aerate the water properly. Frequent flushing of the pond with fresh tidal water compensates the loss of nutrients, maintain variation in the salinity, dissolved oxygen and pH to a minimum.

Pond bottom. The pond bottom is to be checked for any silt deposit. If its level is more than wanted, annual effort for desilting is needed so as to limit the silt level below 20 cm. This would induce the growth of food organisms and bottom fauna to be in proper shape and the prawn may have normal growth. In creating such a situation, there is oftenly abundance of nereid worms and other bottom organisms which are natural sources of food for prawns.

Growth of vegetation along pond margin. For a proper maintenance of stock pond, there is need of growth of marginal vegetation of long stalked grass in the intertidal belt. This plantation will help in the production of periphytic diatoms and accumulation of zooplankton. The presence of these marginal vegetation will provide shelter to the moulting prawn, and also provide shelter to them in the submerged grassy mat.

In some brackish water ponds, farmers implant bundle of dry twigs or palm leaves in different places of the pond bed @ 200 bundles per ha. This measure ensures adequate periphytic growth and protection against hot weather for the prawns.

Feeding of adult prawns. It has been found in most cases of brackish water pond culture that the natural food reserve lasts for first two months for the stocked specimens of prawn. After this period, additional feeding is required in the form of supplementary feed. The stipulated requirement is met with by providing the following items :

(1) Trash fish, molluscan meat and wastes from slaughter house can be given.

(2) Boiled offal cut into small pieces and kept on feeding trays (0.15 m^2) and placed at two corners in a 0.02 ha pond (15 trays with 1 m^2 area each) per ha to be evenly placed. Special trays made of bamboo screen may be kept just submerged underwater and held in position by bamboo poles. The feed is given @ 1 to 2% of the estimated weight of the stock of prawn. This method of providing feed is readily utilized by the growing prawn.

Harvesting Methods

Simplest method of complete harvesting is by dewatering the pond and practising hand picking method to catch hold of prawns. Prawns of all size groups are picked up by this manual method. However, this is possible only in the case of small ponds and in post-winter months when the water level is low. Dewatering of a large sized brackish water pond would be a gigantic task and a non-profitable endeavour due to heavy cost involved on manual labour. Therefore, some other method, such as trapping, is adopted to harvest prawn crop.

Netting. This method is practised for small ponds in any part of the year. The net is, as usual, made up of nylon thread, woven to 1/4″ mesh-work. The foot line is laced with a strong rope which is given weights so that the lower portion of the net is dragged below the silt line at the time of operation. The prawn specimens buried in silt get disturbed and become available for net-catch. In post-winter months, when the water level is low, the netting practice is found comparatively more effective. It has been observed that in most cases as many as 30% of the total crop is collected in the first haul. After a gap of 2-3 days the netting is repeated once or twice or more times as the need be so that maximum quantum of the prawn crop is harvested.

Trapping. In brackish water ponds of big size, neither dewatering device nor netting method of harvesting can be effective for harvesting, nor it would be economical in consideration to the cost involved. Trapping is the best course, as this prawn is of migratory nature and the adult prawns generally move against current. When water enters the pond through the sluice gate with considerable force, the prawns, because of natural habits, congregate around the sluice opening. If the natural current is not sufficient, the current may be generated by pumping water through a small opening in the dyke. For operation, a nylon thread woven net is spread in the area of the sluice opening and the same is lifted periodically and the trapped prawns collected in the bamboo baskets. At the sluice gate opening, cages may be placed skilfully so that the moving prawns may get entrapped. This is possible due to their moving or migratory tendency.

Results of operation generally show that a catch of 10% of the total crop can be collected per hour of such operation and repeated actions may give yield of large number of prawns.

Annual Crop

In most cases, it has been marked that specimen of *P. monodon* grow to 250-275 mm in total length after one year. However, with proper food availability, supplementary feeds and suitable water conditions, the growth attained may be 330 mm in total length. The specimen of 250-270 mm size carry weight of 140-160 gram. The fishery of this prawn is usually done during March to June.

Short term crop. In practising culture of prawn, *P. monodon,* culture for 3 months' duration is also in vogue. If the pond is given due attention in respect to maintenance of water quality, availability of natural food, provision of supplementary feeding, the young ones of 40-50 mm total length generally attain 150-170 mm size (as total length) with an average weight of 28 to 35 gram after rearing in stock pond for 3 months. Generally 28 to 35 numbers of prawn attain 1 kg weight. If the stocking density with young prawn of 40-50 mm total length was 20,000/ha, it has been found that there was chance of 50-55% survival.

Market Values of the Grown-up Prawn

P. monodon has great market value, and is considered as prized item. Its demand is high for domestic consumption as well as in export market. Those in 160-170 mm total length size, having weight above 33 gram (*i.e.*, 25-30 individuals weighing 1 kg), fetch the maximum price in the export market. The actual price range varies from time to time, in respect to rupee values but, a decade before, the range of sale value was 120-150 per kg.

Prawns of small size, *i.e.*, 100-120 mm total length range are also sold like hot cake to the local consumers. 100-150 individual prawns in such size range weigh 1 kg.

25

Shrimp Culture

Shrimps (commonly known as brine shrimp) include several genera of the family *Sergestidae* (Order Decapoda, Suborder Macrura) of Class Crustacea. Shrimps are smaller in size (in average 2.5 cm in length) than prawns (in 5 cm to 30 cm length). The more common species are of Acetes which appear in shoal in the inshore waters (Fig. 25.1). In Indian waters, Atremia forms a very important input for prawn hatchery. It is commonly utilized for feeding the larva of wide variety of organisms including prawn larva. The culture of Atremia is done in high brine with low oxygen level. Due to this unique preferred living condition of Atremia, there is no risk of any predators or other enemies. Its culture has gained popularity for its commercial value as very popular prawn feed. Central as well as state governments of the concerned coastal areas are giving great attention to aquaculture, especially coastal aquaculture practices which includes shrimp culture. Coastal water bodies having large areas are being utilized for culture of shellfish, finfish, prawn culture and shrimp culture. Since large number of fishermen are involved for fulfilment of infrastructural needs, operation of nets, screens, traps, provision of feeds and overall maintenance of the water bodies containing items of culture and, nevertheless, in establishment of great number of hatcheries, the coastal aquaculture is providing grand employment opportunity to the fishermen and their dependents. Out of 1.456 million hectare brackish water area, 0.902 million hectare is being utilized currently for shrimp farming.

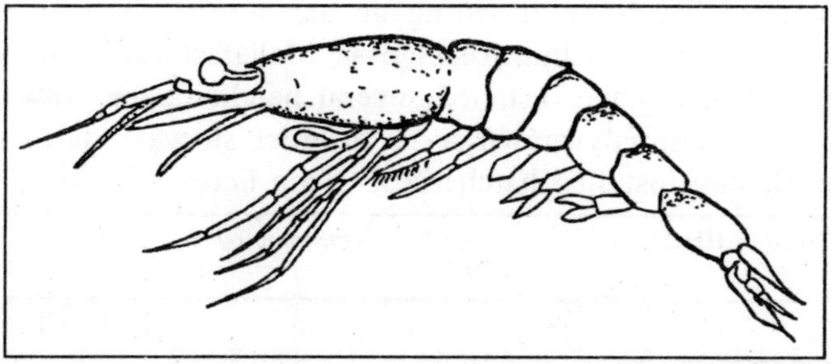

Fig. 25.1. Outline figure of Acetes.

Several big houses like Tatas, Thapars, Hindustan Lever and Godrej have ventured to take up shrimp farming in a big way. Tatas have already set up a Rs. 20 crore project as a joint venture with the state government of Orissa with Technical collaboration from Aquatic Farms Inc. of Hawaii. The joint venture company got the name of Tata Aquatic Farm Orissa Ltd. It began its work by acquiring 700 acres in Orissa. This company expanded functional activities, and is producing export oriented items. After gaining full capacity, this company was expected to generate Rs. 30 crore annually in foreign exchange.

Another project, Thapar Waterways Ltd., which is, likewise, an export oriented project, came up in Nellore district of Andhra Pradesh on the banks of Kendaleru river. This company has installed shrimp farming, to start with, in a 200 acre area. The farm became operational in 1990 and turned to become a well managed export trade. The company got technical collaboration with France Aquaculture Co.

Hindustan Lever Ltd. company has its business activity, related to aquaculture, in West Bengal. This company has developed its own technology at its West Bengal Aquaculture Farm. Now the commercial Production of shrimp is in operation in a big way. Besides, the state government has encouraged the co-operative societies to take up prawn and shrimp culture practices along its coastal waters in planned ways. In this respect, several sorts of facilities are provided to encourage the fish farmers to undertake their prawn and shrimp culture practices in intensive way. Kerala government is not anyway lagging behind in the installation of prawn and shrimp culture farms. An additional involvement has been in setting up hatchery at various places including Kannur, Cochin. Also Orissa, Maharashtra, Tamil Nadu, Andhra Pradesh have installed several hatcheria for prawn and shrimp which supply prawn seeds to brackish water farmers. A list of already installed hatcheria is given here.

Place of Hatcheria	No. of Mill	Ownership by	Varieties of Seeds available
1	2	3	4
TNFDC Hatchery, Tamil Nadu	25	Tamil Nadu State	Prawn, Tiger white prawn
Neelankarai (Tamil Nadu) Regional Hatchery, Trichur.	5	Tamil Nadu State	Tiger white prawn, Shrimp
MATSYAFED Hatchery Mopla Bay, Cannanore Dist., Kerala.	10	Kerala State	White prawn
Prawn Hatchery, Badapokaran, Distt. Thana, Maharashtra.	2	Maharashtra State	Prawn
Ciba Prawn Hatchery Narrakai, Cochin, Kerala.	Experi-mental	KAR	Tiger prawn, White prawn
MPEDA Hatchery, Vallapattam, Cochin, Kerala.	5	MPEDA	Tiger prawn, White prawn
OSPARC (MPEDA), Gopalpore on sea, Ganjam Distt., Orissa.	25	MPEDA	Tiger prawn
TASPARC (MPEDA) Rama Talkies Road, Srinagar, Visakhapatnam Andhra Pradesh.	40	MPEDA	Tiger prawn
Hindustan Lever Prawn Hatchery, Mullukedu, Coveiong, Madras (Chennai).	6	Private	Tiger prawn
George Malio 41, Venkala, Naryana Road, Madras (Chennai).		Private	Tiger prawn
Pioneer Aqua Farms Hatchery, Pereisa Street, Tuticorin.	10	Private	Tiger prawn, White
Victory Aquafarms, Patinamardhur Village, Tuticorin, VOC Dist.	25	Private	Tiger prawn, White
Deejay Research and Breeding Farm	20	Private	Tiger prawn, White
Prabhat Nagar Road, Hanovar.	20	Private	Tiger prawn, White

Hindustan Lever Ltd. company undertook intense innovative research work in the development of shrimp farming. They achieved success in producing 10 tonnes of shrimp in a hectare of water area in two crops per year at their Sudeshkeli brackish water farm in Sundarban area of West Bengal. Their record of production surpassed the highest output achieved by the Chinese brackish water fish farmers where the harvest did not exceed two tonnes per year per hectare. This record level of production rate in West Bengal, can be compared anywhere in the world for excellence.

It is a recognized fact that Orissa is blessed with over 20,000 hectare of brackish water areas which are suitable for development of coastal aquaculture, especially shrimp culture. These are located along their four coastal districts. Besides the state government fishery department, several private companies have ventured in a big way and in Chilka Lake, an aerial view gives an enchanting scenery which probably never was a dream of the central or state government a decade before. The set up of brackish water shrimp culture in Orissa is a development of the present decade. These activities in Orissa and other coastal states have opened up new vista in the direction of improvement of rural economy. Through venture by several private companies, in co-operation by the Orissa government, great progress has been made in the setting up of small shrimp ponds in the marginal areas of Chilka lake. Selected hatcheries have been installed and the production of seeds from these have facilitated undertaking shrimp culture in small ponds connected with the marginal areas of Chilka lake.

One instance of blue revolution in aquaculture development in India is the increase in production of shrimps through coastal aquaculture (Fig. 25.2). This success was achieved owing to the systematic efforts of the central and state government organisations, both in respect to research and development, and also due to the technology transfer from other Asian countries. Whatsoever, India has gained much economic advantage through many aspects of aquaculture practices, but shrimp culture being the most pronounced aspect. Bannot Paul (1932), Jhingran (1969, 1972), Raje and Ranade (1972), Kuttyamma (1975), Biswas (1998) have provided useful informations on various aspects of penaeid prawns, specially about shrimp culture.

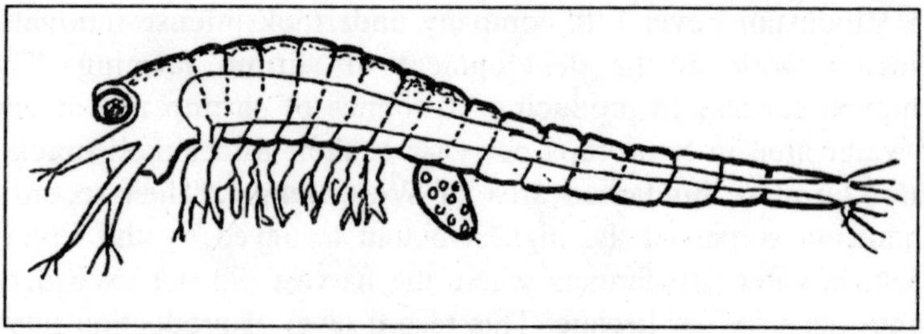

Fig. 25.2. Outline figure of Atremia.

Export of Shrimps

After a gap of 11 years, India in 1997 has emerged as the main supplier of frozen shrimp to Japan, pushing Indonesia to the second position. According to Marine Products Experts Development Authority (MPEDA) sources, the share of Indian shrimp in Japanese market (in terms of value) has increased from 12.20 per cent in 1993 to 21.96 per cent in 1997. Compared to 1996, Indian export of frozer shrimp to Japan had increased considerably during 1997, establishing a record increase of 11.21 per cent in terms of volume and 21.63 per cent in respect to market price value.

Regarding export business in terms of volume, the registered records from MPEDA sources reveal that the exports have increased from 49.823 tonnes to 57.627 tonnes in respect to volume and Rs. 1,653.59 crore (462.54 million US dollars) to Rs. 2064.91 crore (569.63 million US dollars in terms of marketed price during the period. Obviously these figures are most encouraging towards boosting of economy.

Brine Shrimp Culture in Aquarium

Brine Shrimp i.e. *Atremia,* is a very important part of prawn culture as it is used for feeding the larvae of a wide variety of organisms including prawn larvae (Fig. 25.1). As the culture of Atremia typically requires a high brine with very low oxygen level, where predators and other enemies are excluded, this shrimp has its own life of paradise with special environment. Though its culture practice is in vogue in several coastal states where several hatcheria have been installed for the production of seeds of prawn and shrimp, Jamnagar and Kutch districts in

Gujarat are the main areas where atremia has sizable natural population (Fig. 25.3). It has been found that the availability of atremia cyst (which contain eggs) is dependent on the salt level of these areas. The availability is more from March to early June only. However, the installation of big and small hatcheria plants have solved the problem for a continuous supply of seeds. Atremia in a natural condition can be collected from the salt pan of Jamnagar district and put in a suitably designed aquarium with sea water for hatching. It is required that the salinity of seawater be brought down to 28-30 ppm level by adding freshwater in the aquarium. If shifted on rotatory wheel foot to an open area, a full aeration can be provided to the aquarium. During night a bulb of 100 w should be kept hanging above the aquarium. As said before, cysts are produced by all adult specimens after maturation. The cysts contain fertilized eggs. The cysts can be collected by cyst collector and they can be stored in another container having brine water with salinity between 220 to 250.

Fig. 25.3. Scenery of coastal shrimp culture in a estuarine water area.

After sometime the fertilized eggs contained in the cysts hatch out and free swimming nauplii (larva) are produced. For a proper growth of the larva the nauplii are transferred to a cement *cistern*, measuring 1.7 × 0.85 × 0.6 metric, holding 867 litres of aerated sea water. For getting further proper development, artificial feed in the form of boiled rice bran should be provided. Thus, 50 gram of rice bran should be boiled in one litre of freshwater, cooled, macerated and later filtered through 60 micron nylon or boiling silk. The resultant fine suspension should be daily fed to the growing larvae. Self-made "turbidity stick" should be used to check the turbidity level.

The gradual process of development of the larva leads to the emergence of adult atremia, which requires higher salinity. For this, salt pan brine is added at the rate of 5 ppm. While adding the salt pan bine of 5 ppm each day, care must be taken so as the delicate adult atremia do not get adversely affected. After formation of brood sac, salinity should be further raised gradually by adding salt pan bran of 5 ppm each day. Stock of brood should be maintained for further use by cleaning bottom debris of the aquaria at least once in 4 days, as per advice by Kadri *et al.*, 1990.

Shrimp Culture in Ponds

The technology involved is the same in ponds as described for culture in aquaria, but some additional informations of specific kind are needed.

(1) Cysts weighing 1 gram are used to initiate shrimp culture. It contains 1,70,000 bags with 0.25-0.30 mm diameter.

(2) Next step is hatching to adulthood. It is generally seen that hatching starts 24 hours after inoculation and the process is completed in 48 hours. From one gram of cyst as many as 1,02,000 numbers of free swimming larvae come out. The survival percentage of the young one generally is 60 per cent, as reported by Kadri *et al.* (1990) and, earlier, by Dwivedi *et al.* (1979).

(3) The newly hatched nauplius is 0.4 mm. At this stage it can be utilized for feeding prawn larvae, particularly its mysis stage. In view of this prospect, the nauplii of different stages could be collected, and frozen for feeding of the post-larvae of prawns.

(4) Generally 17-19 days are required for the attainment of adulthood by nauplii. From 1 gm cyst generally as many as 20,000 numbers of adult atremia get developed. However, in provision of optimum culture conditions, in less than 15 days duration, larvae can attain the adult form. However, time taken by atremia to mature may be less, *i.e.*, only 12 days, if conditions in water are made favourable in all respects.

The adult atremia develops brood sacs on further rearing for 11 to 13 days. Generally brood sacs are formed with 20 to 67 numbers of cysts in each of them. The findings by several research workers lead to conclude that the number of cysts formed is variable. Weight of brooders along with brood sacs is usually 6 to 7 mg. As all adults have been observed to carry brood pouch, it is apprehended that a parthenogenic condition existed. However controversy remains unsettled as yet, as other ideas have come up, as mentioned ahead.

(5) Experimental results show that 20,000 numbers of Atremia can be maintained in 767 litres of water, which amounts to 23 atremia per litre of water. In outdoor pond culture as many as 100 atremia can be produced per litre of water.

(6) Rice bran is considered as suitable artificial feed for developing atremia. It should be fed at the rate of 0.5 mg per individual (*i.e.* at the rate of 7% of body weight). However, caution is required in the use of fertilizer. If the culture work is in the cement cistern, there should not be any application of organic or inorganic fertilizer before a after inoculation of atremia.

(7) In cultural practices it has been found that salinity extent has great role to play in the population increase. The nauplii and seeds are released by brooders in a salinity range between 80-180%. A closer examination of the adult specimens reveals occurrence of Ovo-Vivipaity. However oviparous atremia was observed in salinity between 150-200%, viviparous condition upto 150% salinity, and ovo-viparity upto 150%, salinity.

(8) Some research works have reported the production of 10 kg. of procured cysts per hectare per month, while others have found potential of atremia eggs about 36 kg per hectare in nature. This figure can go higher by following proper management principles, that is, by undertaking culture of atremia in controlled conditions.

REFERENCES
(For Chapters 22-25)

Arnstein, D.R. and Beard, T.W. 1975. Induced maturation of the prawn, *Penaeus orientalis* kishinouge in the laboratory by means of eye-stalk removal. Aquaculture 5(4); 411-12.

Bannot, Paul 1932. The California Shrimp Industry. Calif. Div. Fish and Game, Bur. Comm. Fish, Fish Bull. 38; 20 pp.

Bauman, R.H. and Jamadre, D.R. 1990. A practical method for determining quality of *Penaeus monodon* (Fabricus) fry for stocking in grow-out ponds. In Proc. Aquatech 90 Conference, Kuala Lumpur, Technical and Economic aspects of Shrimp-farming, New MB, de Sarain H. and Singh, T. (eds). pp 124-31.

Biju Kumar, A. 1998. Aquaculture : A Kerala Model. Yojana March 1998 : 42-43.

Biswas, K.P. 2000. Shrimp Farming — An Ecological and Economic Consideration. Everyman's Science, Vol. XXXIV (4) (Jan-Mar, 2000) : 200-02.

De, D.K. and Subramanyam, M. 1975. Transport of Bagda *Penaeus monodon* Fabricius seed under Oxygen packing. Sci. & Cult. 41(12) : 588-89.

Dwivedi, S.N. 1979. Shipping and Marine Industries Journal, 5(4) : 139-41.

Fielder, D.R., Greenwood, J.G. and Ryall, J.C. 1975. Larval development of the tiger prawn, *Penaeus* esculentus Haswell 1879 (Decapoda, Penacidae) reared in the Laboratory. Aust. J. Mar. Freshwat Res. 26(2) : 155-75.

Ganapati, P.N. and Subramanyam, M. 1964. The prawn fishery in Godavari Estuary J. Zool. Soc. India 16 (1 and 2) : 11-20.

George, M.J. 1962. On the breeding of penaeids and the recruitment of their postlarval into the back waters of Cochin. Ind. J. Fish, 9A(1); 110-16.

——. 1966. On a collection of penaeid prawns from the offshore waters of the south-west coast of India. Proc. Symposium on Crustacea 1 : 337-46.

Gilda Lio PO 1991. Diseases of Pond culture shrimp with emphasis on prevention strategies. Aqua cult. Deptt. South East Asia Fisheries De V Centre, Philippines.

Gopalkrishnan and Mallikarjun Rao, R. 1969. Development of prawn fisheries in Indian Reservoirs. Proc. Seminar on the Ecology and fisheries of Freshwater Reservoirs. Barrackpore, Nov. 27-29 : 551-60.

Holder, D.D. 1978. Induced maturation and breeding of *Penaeus monodon* under brackishwater pond conditions by eyestalk ablation Aquaculture 15 (2) : 171-74.

Ibrahim, K.H. 1962. Observations on the fishery and biology of the freshwater prawn *Macrobrachium malcolmsonii* Milne Edwards of River Godavari. Ind. J. Fish A 9(2) : 433-67.

Janardana Rao, K. 1973. Observations on the seasonal abundance of post larvae of Metapenaeus (Fabricus) and *M. dobsoni* (Miers) in the Pulicat lake. Ind. J. Anim. Sci. 43 (10) : 948-55.

——, Venkateswara Rao, K., Ramaraju, T.S., Subha Rao, K., Ramkrishna Rao, D. and Bose, P.S.C. 1979. Observations on the freshwater prawn and seed resources of the Godavari Western Delta, Andhra Pradesh. In : Symposium on Inland Aquaculture (Abstracts) February 12-14 : 61 CIFRI, Barrackpore.

Jhingran, V.G. *et al.*, 1963. Report on the fisheries of Chilka Lake, 1957. Bull. cent. Inland Fish Res. Inst. Barrackpore, No. 1 : 113 p. (Mimeo.).

—— and Natrajan, A.V. 1969. A study of the fisheries and fish populations of the Chilka lake during the period of 1957-1965. J. Inland Fish Soc. India 1 : 49-126.

——. 1969. Problems of coastal aquaculture in India J. mar. biol. Ass. Ind. 11 (1-2) : 56-61.

——. 1972 Role of Chilka lake fisheries in the development of coastal aquaculture in Eastern India. In : T.V.R. Pillay (ed.). Coastal Aquaculture in the Indo-Pacific Region. 109-15. London : Fishing News (Books) Ltd.

Kadri, S.A., Trivedi, C.R., Patil, M.I. and Charya, N.D. 1990. Culture of brine shrimp in controlled conditions. Fish ohyme, 10(9); 36-40.

Kewalramani, H.G., Sankholi, H.G. and Shenoy, S.S. 1971. On the larval history of *Macrobrachium malcolmsonil* (H. Milne Edwards) in captivity. J. Ind. Fish Ass., 1(1) : 1-25.

Krishnayya, Ch. G. and Rao, K.J. 1975. Observations on the qualitative and quantitative abundance and ingress of penaeid post-larvae in Pulicat lake. Indian J. Anim. Sci. 45(6) : 397-403.

Kurata, H. and Shegueno K. 1976. Recent Progress in the farming of penaeid shrimp. FAO Tech Conf. Aquacult. Kyoto, Japan, 26 May – 2 June 1976. FIR : AQ/ Conf/76/R17 : iii + 24 p (mimeo.).

Kuttyamma, V.J. 1975. Studies on the relative abundance and seasonal variations in the occurrence of the post-larvae of three species of penaeid prawns in the Cochin back-waters. Bull. Dept. Mar. Sci. Univ. Cochin 7 (1) : 213-19.

Lin Kwei, C. 1991. Prawn culture in Taiwan, College of fisheries, Institute of Aquaculture, Philippines.

Ling, S.W. 1962. Studies on the rearing of larvae and Juveniles and culturing of adults of *Macrobrachium rosenbergii* (de Man). Tech. Rep. Indo-Pacif Fish Coun. (57) : 15 p.

——. 1964. A general account of the biology of the giant freshwater prawn *Macrobrachium rosenbergii* and methods for its rearing and culturing. Paper presented in the Indo-Pacif. Fish Coun. 11 : 4.

——. 1969a. Method of rearing and culturing *Macrobrachium rosenbergii* (de Man) FAO Fish Rep. (44) 3 : 291-309.

—— and Costello, T.J. 1976. Review of culture of freshwater prawns. FAO Tech. Conf. Aquacult., Kyoto, Japan, 26 May–2 June 1976. FIR : AQ/ Conf/76/R. 29 iii + 12 p. (mimeo.).

Matida, Y. 1967. The role of soil in fish pond productivity in Asia and the Far East. FAO Fish Rep. (44) 3 : 54-63.

Meeran, N.K. and Sebastian , M.J. 1973. On rearing the larvae of the giant freshwater prawn, *Macrobrachium rosenbergii* (de Man). In : Proc. Seminar on Mariculture and Mechanised Fishing, 28-29 November 1972 : 94-96.

Mohamed, K.H. 1969. Trends in crustacean fisheries. Indian Fing. 19(9) : 36-40.

Mohammad, K.H. 1967C. FAO World Sci. Conf. on Biol. and culture of shrimps and prawns. Mexico FR/BSCP/SSM.

Panouse, J. and Menon, M.K. 1956. Prawn Fisheries of India. Proc. Indo-Pacific Fish Coun. 6(3) : 328-44.

—— and Aiyar, R.G. 1939. The brackishwater fauna of Madras. Proc. Indian Acad. Sci. 6B : 284-337.

——. 1943. Eye stalk and ovarian growth in crustaceans. G.R. Acad. Sci. 217 : 553-55.

Rajyalakshmi, T. 1960. Observations on the embryonic and larval development of some estuarine palaemonid prawns. Proc. nat. Inst. Sci. Ind. 26B (6) : 395-408.

——. 1964. On the age and growth of some estuarine prawns. Proc. Indo-Pacif. Fish Coun., 11th. Sess 11 : 52-83.

——. 1974. The freshwater prawn *Macrobrachium malcolmsonii* for use in pond culture. Seafd. Export J. 6 (3) : 25-32.

Raman, K. 1964. On the location of a nursery ground of the giant freshwater prawn, *Macrobrachium rosenbergii* (de Man). Curr. Sci. 33(1) : 27-28.

——. 1967. Observations on the fishery and biology of the giant freshwater prawn, *Macrobrachium rosenbergii* (de Man). Proc. Symp. Crustacea Mar Biol. Ass. India Part 11 : 649-69.

Ramkrishnaiah, M. 1979. Observations on Post-larval incursions and fishery of penaeid prawns in Chilka lake. J. Inland Fish Soc. India Vol. II, No. 2 : 31-40.

Ranade, M.R. 1972. Propagation of freshwater prawn. FAO Aquacult. Bull. 4(2) : 4-5.

Rao, K.J. Krishnayya, Ch. 6. 1974. Penaeid prawn catches from Pulicat lake in relation to ingress of post larvae and lake hydrography. India Indian J. Fish 21(2) : 445-53.

Rao P. Vedevyass and Kathirvel, M. 1973. On the breeding of a penaeid prawn, *Metapenaeus dobsoni* in the brackish water medium. Ind. J. Fish 20 (1) : 228-30.

Tsai, Chia Kuang. 1991. Water quality Management. College of Fisheries, Institute of Aquaculture, Philippines.

Santiago, Jr. A.C. and Wear, R.G. 1976. Induction of maturity and spawning in *Penaeus monodon* Fabricus 1978 by unilateral eyestalk ablation (Decapoda, Natantia). Crustaceana 31(2) : 218-20.

Subramanyem, M. 1967. Fluctuations in the prawn landings in Chilka lake. Proc. Indo-Pacific Fish Coun. 12(2) : 203-17.

—— and Rao, K.J. 1968. Observation and on the post-larvae prawns (Penaeidae) in the Pulicat lake with note on their utilization in capture and culture fisheries. *Ibid.* 13(2) : 113-27.

26

Cold Water Fisheries

The mountainous areas of Himalayas and some other places of South India are adorned with a large number of natural lakes, streams and natural ponds which have perennial water supply provisions and water bodies have low temperature, ranging from 10 to 20°C, high oxygen content, low level of carbon dioxide and fast running water. The rivers and streams of adjoining hills are shallow with rocky bottom, clear water and plenty of natural food in the form of vegetation and algal matters. The ecological condition, particularly physico-chemical characteristics is such that the fresh water fishes of the plain areas cannot survive long. However, several fishes with adaptability to survive in cold water conditions have paradise life in the hilly areas of Himalayas and Nilgiri areas of South India. Thus *Salmo* sp. (trout) *Oncorhynchus* (trout), mirror carp (*Cyprinus* sp.), tench (*Tinca tinca*), golden carp (*Carassius auratus*), Mahseer (*mahseer* or *Tor* sp.), snow trout (*Schizothonax* sp.), *Barilius, Labeo* sp., *Garra* sp., *Glyptothorax* sp. and few more small sized fishes of little economic value are found in various cold water sources of Jammu, Kashmir, Himachal Pradesh, Assam, Manipur, Sikkim, Nilgiris, Kodai hills (Tamil Nadu), Travancore (Kerala) areas of India and also in Bhutan, Nepal etc. Several fishes occur naturally in the cold waters and are adapted to enjoy cold water habitat. However, several species of trout such as the rainbow trout (*Salmo gairdneri*), the brown trout (*Salmo trutta*), the brook trout (*Salvelinus fontinalis*), the cut throat trout (*Salmo clarke*), the golden trout (*Salmo aquabonita*), gunapee trout (*Salvelinus aureolus*), which are exotic in nature, but now well settled in the cold water areas of India, carry special significance for tourists for their sporting use as game fishes and considered as source of revenue to the state concerned.

General Account

Fisheries of the cold water habitat are described here under 10 heads: 1. the brown trout or European trout (*Salmo trutta*), (ii) the rainbow trout (*Salmo gairdneri*), (iii) the brook trout (*Salvelinus fontinalis*), (iv) the Mahseer (*Tor tor, T. Putitora*), (v) the mirror carp (*Cyprinus carpio*), (vi) the tench (*Tinca tinca*), (vii) the crucian carp or the golden carp (*Carassius carassius*), (viii) the snow trout (*Schizothorax* sp. and *Schizothoraichthys* sp.), (ix) the Indian hill trout (*Barilius* sp.) and (x) the carp (*Labeo dero*; *L. dyocheilus, L. gonius*). Besides these, *Garra* sp., and *Glyptothorax* sps. are considered as cold water fishes. Jhingran and Sehgal (1978) and Tilak Raj (1982) have provided useful informations about cold water fishes, game fishes, sport fisheries, etc.

1. Brown Trout (*Salmo trutta*)

This fish was introduced in India from Europe during the turn of the present century in North West Himalayas (Fig. 17.1). The fish has become almost endemic in the streams and lakes of Kashmir and Himachal Pradesh but lesser in Uttarakhand. Although the fish is not as sporting as other species of trouts and salmon, it is still very popular as a game fish because of its abundance in places of tourists attraction. In nature brown trout feeds more regularly at the surface than the rainbow trout. It takes several species of insects, worms, spiders, snails, frogs, even mice and young birds, if they happen to fall into the water.

2. The Rainbow Trout (*Salmo gairdneri*)

This variety of rainbow trout, like the brown trout, has been transplanted from Europe in Kashmir, Himachal Pradesh, Meghalaya and Nagaland (Fig. 17.2). In Peninsular India, the shasta and irideus varieties, transplanted from Sri Lanka, are available in the Munnar High Range, Nilgiris, Kodai Hills and Kamengundi hills. In 1964-65, golden variety of rainbow trout from Japan was introduced in the Nilgiris and is now available in all the reservoir of the Nilgiris. The steelhead rainbow trout failed to establish in the streams of Kashmir and Himachal Pradesh, but is amenable to artificial propagation in trout culture farms of the two states.

The fish possesses and displays all the qualities of a game fish, with a robust built body, the fish combines great strength with a willingness to fight against any odd in life. Thus after getting hooked, it breaks water like an explosion. It strikes vigorously and does not give up fighting as long as the fins function.

The feeding habits of rainbow trout are quite uniform. They show preference for mature stage of insects, minnows and crustaceans. The fish has so wide choice of food that sometimes even pieces of aquatic weeds, moss, small gravel and even piecc of woods are found in its stomach. The fish enjoys life in running water. It is found in the fast current under overhanging banks or willow clump or in a fast riffle. It avoids passing to sluggish part of the streams. In the undercurrent along the stream banks, where water runs swiftly, the fish gets easily hooked. The fish equally enjoys life in deep reservoirs and natural lakes of southern India. The fish does not feel otherwise in well aerated cold and warm waters (upto 25°C surface temperature of lakes and reservoirs in Munnar High Range and Nilgiris) of South India. They take this habitat as congenial to their needs of life.

3. The Eastern Brook Trout

This fish (e.g., *Salvelinus fontinalis*) is another example of cold water fish. It was very much popular fish when introduced in Indian waters in 1919. It was later transplanted from Kashmir to Kullu waters of Himachal Pradesh. As it showed poor adaptability, it is loosing the status of high demand. Because of its poor tolerance to temperature, it is difficult to culture this fish. On the other hand, the brook trout of the Northern Main Farm pond has better adaptability. The fishes spawn successfully in the springs or in the tiny spring tributaries which drain into the ponds.

The eastern brook trout of the cold water ponds has fastest growth rate among the three species and has most palatable flesh. But it is difficult to culture this trout in ponds due to poor adaptability, as mentioned before. Moreover, this fish has the disadvantage of susceptibility to diseases like fin-rot and furunculosis and also to pollutants and toxicants of various kinds.

4. Mahseer

It is most popular and a mighty game fish of India and is called Mahseer due to its big head. Several species of the fish are found in the mountainous streams and rivers of India. *Tor masal*, *Tor khudree*, *Tor mussulate* are well-known species but *Tor putitora* is the common mahseer occurring in the rivers and streams from Kashmir to Assam (Fig. 26.1). *Tor tor* is next well-known cold water species, resembling *T. putitora* in many respects (Fig. 26.2). In Kumaon, *Tor putitora* is commonly found in Naini Tal, Bhim Tal, Sat Tal etc. *Tor mossal* inhabits the rivers of Kashmir, Sikkim and Assam.

The mahseer is primarily an inhabitant of the rocky mountaineous parts of the perennial streams of India. However, they are not available in areas where the river gets broader and shallower and have a sandy bed. They are abundantly found in deep still pool without a rock. The best fishing spots are the rocky mountain part of the stream. The fish congregate in deep runs specially where a fall enters a pool. They are found in large numbers also in the eddies of these runs.

The fish grows to about 1.5 m in length when fully adult, and breeds from July to September. The fish feeds on hard fishes likes *Garra gotyla*, Young of *Chela argentea*, *Channa gachua* etc. It has preference for any particular species of fish whichever is hardiest.

5. Mirror Carp

The mirror carp has three species with specific characteristics in morphological features and behaviour. The species are :

(i) The scale carp, *Cyprinus carpio var communis*.

(ii) The minor carp, *Cyprinus carpio var specularis*.

(iii) The leather carp, *Cyprinus carpio var nudus*.

The mirror carp was first introduced into the upland waters of Ceylone from Russia in 1914. From there it was brought to the Nilgiris in 1939 and stocked in the Ootakmund lake. Another strain was brought to India in 1957 from Bangkok. This strain showed successful result, as the breeding occurred in plains where the temperatures were considerably higher than on the hills. In recent decades, the Prussian strain, *Cyprinus carpio var specularis* has acclimatised and now it breeds freely at various places in the plains in India. The mirror carps were later transplanted from Nilgiris to various places of India including Kumaon in 1947, other place of Himachal Pradesh in 1955, Kashmir in 1956, Sikkim in 1957, Punjab in 1958, Madhya Pradesh in 1959, Bihar in 1960 and Rajasthan in 1960. With the introduction of minor carp to Himachal Pradesh, the state has become self-sufficient in fish production. Nevertheless, this state is now supplying seeds and young fish for stocking to other states of India.

Mirror carp is an ideal fish for introduction in hilly regions where it enjoys life and breeds (Fig. 26.3). This fish has been found to breed in plains as well. The seed is now available in the hilly and many plain water of India. The mirror carp is an omnivorous fast growing fish and is cultured in India along with other major carp.

6. Tench (*Tinca tinca*)

This fish, commonly known as tench, is an exotic fish from England and thrives in the cold waters of Nilgiri hills (Fig. 26.4). It was first introduced by Maclvov in 1874 in Ooty lake. Later the fish was brought to ponds and lakes of Nilgiri hills where it had successful life activities. It has normal activities in cold waters. Later this fish got acclimatisation in Sunkesula Fish Farm, Andhra Pradesh, and afterwards to various plains of India.

It is a bottom feeder consuming insects, vegetable debris and plankton. It grows to a maximum of 50 cm and the recorded maximum weight from Ooty lake is 1.4 kg. It is a prolific breeder and spawning time in Nilgiris has been marked to be February to April. The eggs are adhesive, getting attached to the leaves of aquatic plants. The incubation period lasts for 5 to 6 days.

7. The Crucian Carp/The Golden Carp

The crucian carp, *Carassius auratus*, also known as golden carp, was introduced by Maclvov in 1874 from the native place, Central Europe, in the Ooty lake along with tench. It was later brought to other places of Nilgiris waters and it was found that it got acclimatized in new habitats. Its growth to a maximum

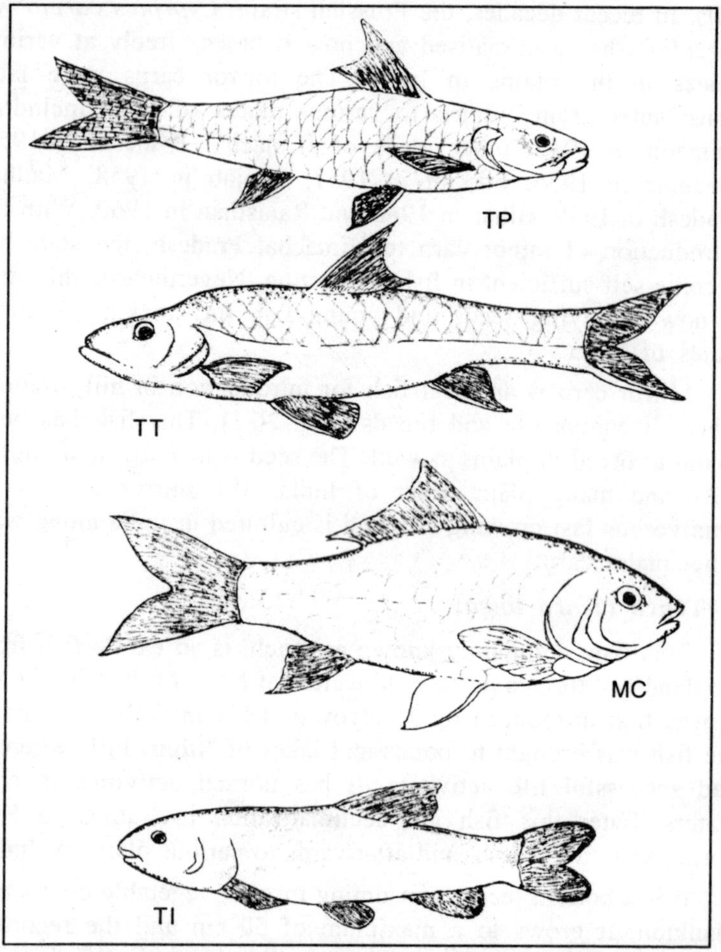

Fig. 26.1 - 26.4. Some common cold water fishes : Mahseer e.g. *Tor putitora* (TP), *Tor tor* (TT), mirror carp, *Cyprinus carpio* var *specularis* (MC), tench, *Tinca tinca* (TI).

length of 4.5 cm or a weight of 1.4 kg was recorded in Ooty lake (Fig. 26.5). It feeds on insects and crustaceans but copepods and cladocerans (crustacean) are preferred food items. It breeds in ponds all the year round. The eggs are adhesive, like those of tench and are attached, on release, to the submerged water plants. Attempts were made to culture the fish in the plains but little success was obtained. Along with the tench, this fish has been bred in the Sunkesula Fish Farm, Andhra Pradesh.

7. The Snow Trout

The snow trout comprise two genera :

(i) *Schizothorax* Heckel, which has a suctorial lower lip,

(ii) *Schizothoraichthys* Misra, which is without a suctorial lower lip.

Another form of Schizothorax, *Schizothorax richardsona* has a suctorial mouth and about 97 scales on the lateral line. The species attains a maximum length of 60 cm (Fig. 26.6). It is found in snow fed streams of Afghanistan, Kashmir, Punjab, Nepal, Bhutan, Sikkim, Assam and other places of eastern Himalayas. In the streams of Himachal Pradesh at high altitude it is abundantly found. The fish prefers to enjoy life in rapids and pools of big terrestrial streams having a temperature range of 8° to 22°C. The fish breeds from March to June in water temperature of 14°C to 21°C. Eggs are laid among gravel towards the bank of the stream in clear water of moderate current at a depth of 30-56 cm. Newly hatched out larvae are 11.4 mm in length with oval yolk sac. Fries possess adhesive apparatus and look blackish in colour.

9. The Indian Hill Trout

This fish is represented by *Barilius bendelisis*, *B. bola* and *B. gatensis*. The *Barilius bendelesis* is an important game fish, and is abundantly found in Assam, West Bengal, Punjab, Simla, Uttar Pradesh, Bihar, Coimbatore, Palghat and Nilgiris (Fig. 26.7). As its another variety, *B. bendelesis* var *chedra* occurs in the northern tributaries of river Beas and inhabits streams having crystal clear water. It prefers rapids as well as pools at high altitude. The fish prefers water temperature between 10-24°C.

B. bola is another species of *Barilius*. The species is called the Indian trout or hill trout and occurs in North-Eastern India down to Orissa. It grows a maximum length of 30 cm and is considered as the chief game fish of Assam rivers. On the other hand, *B. gatensis* is found in Peninsular India, particularly in Coorg, Mysore, Western Ghats, Travancore-Cochin, Malabar and Nilgiris. It attains a maximum length of 15 cm.

10. Carps

Some species of carps, such as *Labeo dero, Labeo dyocheilus*, *Labeo gonius* are counted among the cold water fishes as they occur naturally in cold water of Himalayan region. *Labeo dero* is found in Assam, Darjeeling, Eastern Himalayas, Punjab, Uttar Pradesh and Western Himalayas (Fig. 26.8). They are abundantly found in Punjab and Himachal Pradesh and form an important fishery in the Bhakra reservoir. Like any other cold water fish, it prefers rapids and pool and its collection is generally found associated with mahseer, *Tor putitora*.

Labeo dyocheilus is another carp which is naturally occurring in cold waters. This fish is commonly found in Assam, North Bengal, Nepal, Himalayas and hilly region of Himachal Pradesh. It has been observed that this fish migrates in Kangra valley streams along with mahseer. *Tor putitora*, towards higher altitude during monsoon.

Besides 10 fish groups mentioned above, there are several other fishes of less economic value which are naturally occurring in the cold waters of Himalayan streams. *Garra lamta, G. gotyla* and *Glyptothorax stoliczkal, G. pectinopterus* and *Pseudoecheneis* are few such examples (Fig. 26.9).

It is evident from the yield obtained and demand prevalent in the market that most important species of fish cultured in cold water is the trout. From encouraging results of culture of brown trout (*Salmo trutta*) and rainbow trout (*Salmo gairdneri*) in Kashmir. Fisherics Department of Jammu and Kashmir imported eyed ova of eastern brook trout (*Salvelinus fontinalis*) from Canada and the landlocked Salmon (*Salmo salar*) from the United States of America and the efforts proved successful as the fish got adapted to the cold water conditions of Jammu and Kashmir.

Rainbow trouts, *Salmo gairdneri irideus* and *Salmo gairdneri* Shasta were successfully transplanted in Nilgiris, Kodai Hills (Tamil Nadu), Munnar High Ranges (Kerala). Recent event of transplantation of Japanese strain of *Salmo gairdneri*, golden strain of *Salmo gairdneri* and Kokane salmon (*Oncorhynchus nerka*) have been in waters of Nilgiris in Tamil Nadu.

Trout in Himachal Pradesh was first introduced in river Beas in Kullu valley in 1909 by Howell. This stock got established there, and became source of trout fry in other streams of river Beas, then to stream of Uhl valley, streams of Pabar and Baspa. All above mentioned streams are snow fed streams at about 1600 m. So they proved to be ideal places for the trout fishery.

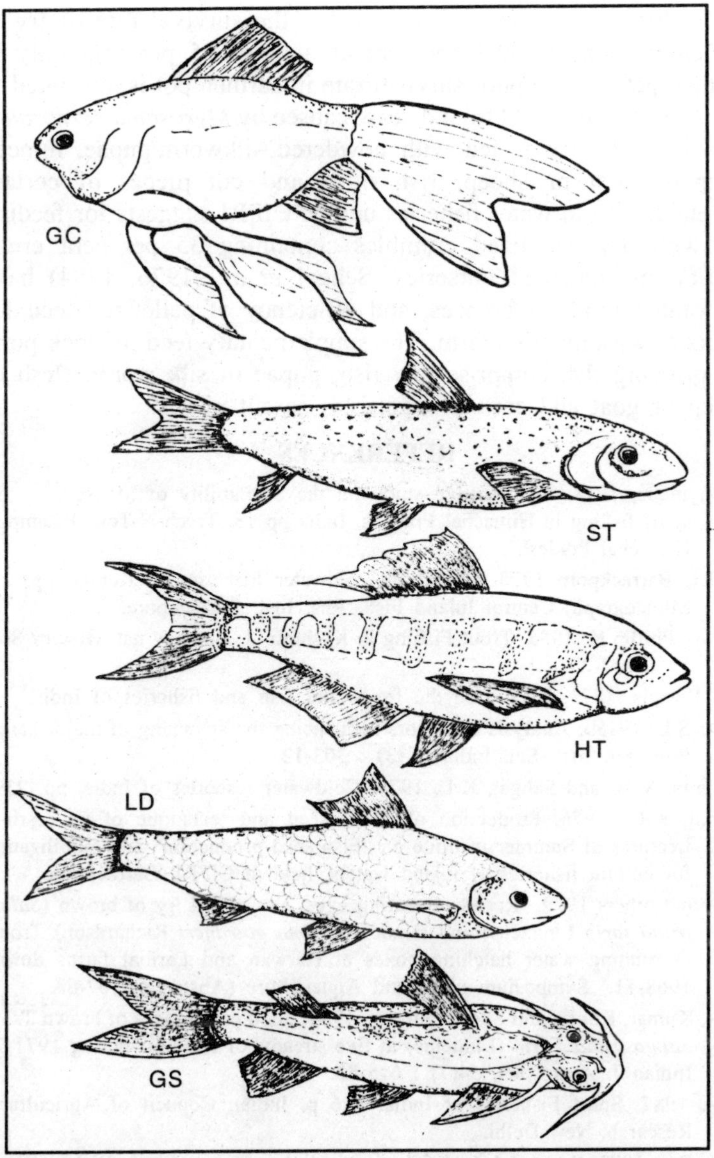

Fig. 26.5. to 26.9. Additional common cold water fishes: Golden carp, *Carassius auratus* (GC); snow trout, *Schizothorax heckel* (ST), Indian hill trout, *Barilius bendelisis* (HT), Indian carp, *Labeo dero* (LD), and Indian hillstream fish, *Glyptothorax stoliczkal* (GS).

In Kashmir it has been observed that the survival rate of fry in cemented tanks is 59.5 per cent as against 7.5 per cent only in earthen ponds. The poor survival rate in earthen pot is attributed to occurrence of dreadful trout disease caused by *Myxosoma cerebrates*. Generally the fry is fed with powdered silkworm/pupae, minced liver of goat or sheep, fish meal and cut pieces of certain vegetables. Cold water fisheries unit of CIFRI suggests for feeding fry with dry pelletized crumbles containing 35 per cent crude portion in cemented nurseries. Sehgal *et al.* (1976, 1984) have elaborated food preferences, and efficiency of pelletized feed for trouts of Kashmir fish farm. The supplementary feed in stock pond for growing fish comprise raw fish, pupae of silk worm; flesh of sheep or goat and certain vegetables as cut pieces.

REFERENCES

Anonymous, 1978. Investigative study on the availability of Mahseer — for sport fishing in Himachal Pradesh, India pp 18. Trech-N-Tour Palampur, Himachal Pradesh.

CIFRI, Barrackpore 1970. Report on coldwater Fisheries in Kerala. pp. 12. Mimeograph. Central Inland Fish. Res. Inst. Barrackpore.

Crowe, Phillip K. 1955. Trout Fishing in Kashmir. J. Bombay nat. History Soc. 53 : 217-28.

Day Francis. 1873. Report on the freshwater fish and fisheries of India.

Hora, S.L. 1945b. Analysis of factors influencing the spawning of major carps. Proc nat. Inst. Sci. India II (3) : 303-12.

Jhingran, V.G. and Sehgal, K.L. 1978. Coldwater fesheries of India. pp. 239.

Sehgal, K.L., 1976. Production of Trout seed and technique of its rearing. Lectures of Summer institute on "Fish seed production" and mobilization for culture fisheries of inland waters, held at CIFRI, Barrackpore.

—— and others 1979. 'Result of rearing eggs, alevin and fry of brown (*Salmo trutta fario* Linnaeus) and rainbow (*Salmo gairdneri* Richardson) Trouts in running water hatching boxes at Harwan and Laribal farms during 1968-71.' Symposium on Inland Aquaculture (Abstr.) pp. 67-68.

——, Kumar, Kuldip and Sundar Shyam. 1984. Food preferences of brown Trout *Salmo trutta Fario* (Linnelus) in five streams of Kashmir during 1971-74 Indian J. Arum. sci. 54(7) : 675-82.

——. 1987. Sport Fisheries of India. 126 p. Indian Council of Agricultural Research, New Delhi.

Tilak Raj, 1982. Sport fish and fisheries in India. Mohan Singh Bishen Singh Publishers, Dehradun, pp. 259.

Walkabout, The 1978. Fish the Himalayas. October 9.

Welborne, J.E. 1879. Poongah (Assamese) Fishing in rivers. Asian, 22 July and 2 September.

27

Riverine Hydrobiological Features

Knowledge of quality of water and plankton biomass in an ecosystem *e.g.* riverine water, is of fundamental importance, as growth of aquatic biota in general and fishes in particular is directly related to these two factors and these are closely involved in the food cycle of fishes and other organisms. For a proper growth of plankton biomass, proper quality of water is an essentiality. As the volume of planktonic organisms is more important than their numbers in utilization of natural water for fish production, a thorough knowledge of the abundance of plankton in time and space in relation to environmental condition has become a prerequisite.

One can observe the occurrence of fluctuation, both seasonal as well as diurnal, in physico-chemical conditions of riverine water and also the effect of the environmental conditions on the plankton population. On the other hand, various king of chemicals or effluents discharged from factories, mills or waste chemicals from sprayed or manured crop fields are swept through rain water or irrigated water to the drains and lanes which form linkage with rivulets or rivers. The fish and fish pond organisms living in such uncertain environment have to sustain the adversities to varying degrees. Temperature, pH, hardness, levels of dissolved oxygen and free carbon dioxide, total alkalinity, phosphate and nitrate are the main fluctuating physico-chemical factors which effect the planktonic population both quantitatively and qualitatively. In general, the hydrobiological features of the pre-monsoon, monsoon, post-monsoon and dry season are diversified. Due to these uncertain situations noticeable diversifications are observed noticeable in the planktonic mass of the rivers and quantum of fish production per hectare from different river sources is varied. These revelations are evidenced from the figure of fish production from different rivers as mentioned ahead in the chapter 'Riverine Fisheries'.

When discussing about the riverine hydrobiological features of India, the main river systems worthy of mention are the following (Fig. 27.1) :

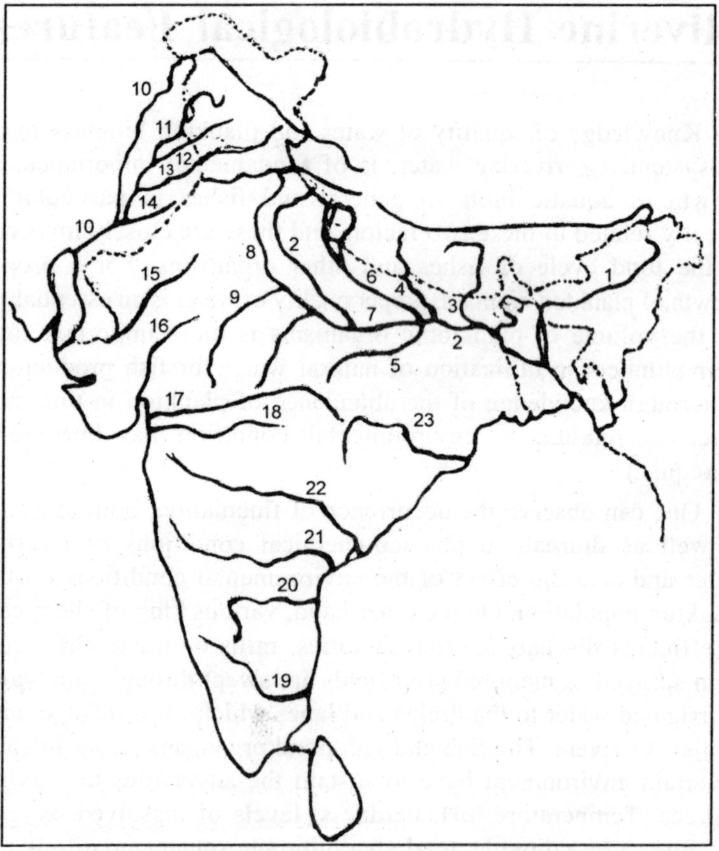

Fig. 27.1. Sketch map of India showing main rivers : Brahmaputra (1), Ganga (2), Kosi (3), Gandak (4), Sone (5), Ghaghra (6), Gomti (7), Yamuna (8), Chambal (9), Indus (10), Jhelum (11), Chenab (12), Ravi (13), Sutlej (14), Luni (15), Sabarmati (16), Narmada (17), Tapti (18), Kaveri (19), Penneru (20), Krishna (21), Godavari (22), Mahanadi (23).

1. The Ganga River System.
2. The Brahmaputra River System.
3. The Indus River System.
4. The East Coast River System : Mahananda, Godavari, Krishna and Kavery.
5. The West Coast River System : Narmada and Tapti.

1. Ganga River System

Ganga River System has the total length of about 8,047 kms. and is considered as one of the largest river systems of the world. It emerges from southern part of Himalayas, drains the southern slopes of Himalayas and passes through the states of Haryana, Uttar Pradesh, Bihar and Bengal (Fig. 27.2). Its tributaries also pass through parts of Rajasthan and Madhya Pradesh.

The tributaries, joining the Ganga in its course through the plain include Yamuna, Gomti, Ghaghra, Gandak and Kosi. The

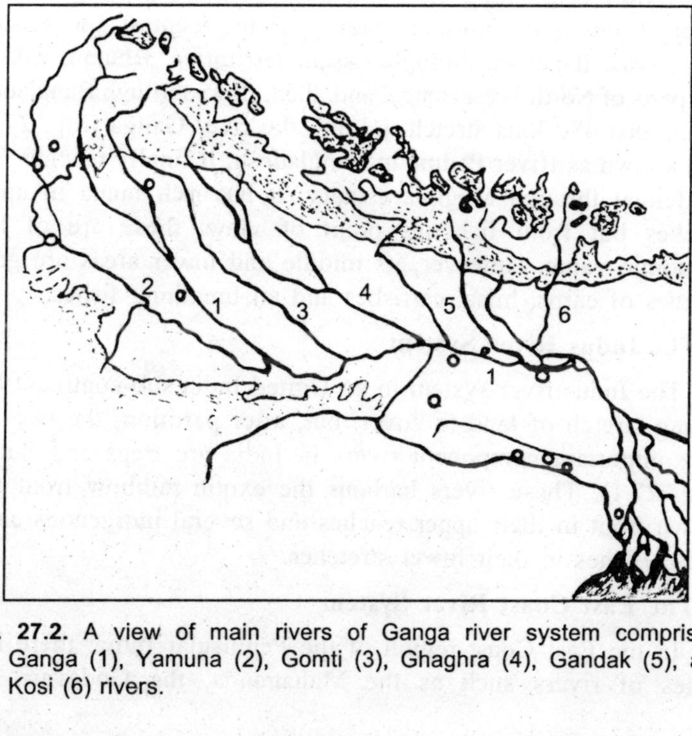

Fig. 27.2. A view of main rivers of Ganga river system comprising Ganga (1), Yamuna (2), Gomti (3), Ghaghra (4), Gandak (5), and Kosi (6) rivers.

Yamuna is the most important tributary and joins the Ganga on its right Bank at Prayag (Allahabad). From Allahabad the river flows eastward for over 640 km. Gomti, Ghaghra, Gandak and Kosi are the main tributaries joining in the left bank.

At the eastern side, the river enters Bengal and turns south. At about 350 kms up from the sea (Bay of Bengal), the Ganga begins to spread into several small tributaries, forming the famous Sundarban delta. Hooghly is a famous distributary and on its bank stands Calcutta. The min channel of Ganga flows in a south-easterly direction through Bangladesh and is then called Meghna river. Plain stream of Brahmaputra joins Meghna and meets the Bay of Bengal. The main Ganga river as well as its tributaries and distributaries are of great sources of fishes, particularly for major carps and famous hilsa.

2. The Brahmaputra River System

The Brahmaputra river system has a total length of 2,900 kms. After originating from Mansarovar lake, passing through Tibet, draining the northern slopes of the Central and Eastern Himalayas, it passes through Assam territories, Bhutan, Sikkim and parts of North West Bengal and, then, flows through Bangladesh for its last 480 kms stretch, joining the river Ganga (Fig. 27.1). It is known as River Padma in Bangladesh. It finally reaches Bay of Bengal through Meghna estuary. It has rich fauna in upper reaches but from fisheries point of view, these are of less economic value. However, its middle and lower areas are good sources of carps, hilsa, catfishes and air-breathing fishes.

3. The Indus River System

The Indus river system in an United India sub-continent had a long stretch of land to cover, but, after partition, the two left over important component rivers in India are Beas and Sutlej. (Fig. 27.1). These rivers harbour the exotic rainbow trout and brown trout in their upper reaches and several indigenous carps and catfishes in their lower stretches.

4. The East Coast River System

In the East Coast region of the Peninsular India, there is a series of rivers such as the Mahananda, the Godavari, the

Krishna and the Kavery (Fig. 27.1). These together cover a stretch of about 6,437 kms. These rivers drain the land areas of the Peninsula, having extension towards west in the east of western Ghats, and towards southern parts of Central India including the Chhotanagpur Hill ranges towards the north-east.

The Mahananda river harbour all the common kinds of Indian major carps that are found in the Ganga System. Other rivers of East Coast Systems harbour several carps, catfishes, murrels, perches, prawns and some other indigenous species. Several species of fishes from the Ganga river system has been transplanted in these peninsular rivers and their flourishing population after establishment in these new environment has enriched the prospects of fisheries from these rivers. A significant aspect is that tributaries of Cauvery river, originating from the Nilgiri hills tract, have flourishing population of coldwater fishes like trout and tench.

5. The West Coast River System

Major rivers of western coast are Narmada and Tapti (Fig. 27.1). The former is 1,312 km long, rises on the Amarkantak Plateau in the Maikala Range from a spring at about 1,060 m height. It flows generally west, then south-west and, afterwards, falls down from a height of 15 metre, forming the Marble Falls. It finally enters Gulf of Cambay by a broad estuary.

Tapti river rises near Multai on the Satpuda Range and flows generally west past Burhanpur and Surat (Gujarat State) to the Gulf of Cambay. The Parna, Girna and Panghara are its main tributaries. The Tapti River covers a total distance of 724 km. Both rivers have rich representation of fauna and flora.

Hydrobiological Features

Most intensive studies have been made in the water condition in 1090 km stretch of the Ganga, extending from Kanpur in the west to Rajmahal in the east. A gross impression is drawn on the basis of many reportings as given below.

Turbidity. It is 1100-2170 ppm during January-June. The turbidity level declines sharply during post-monsoon period, reaching less than 100 ppm by January.

pH. It ranges between 7.45-5.30. It is below 80 during June-August and monsoon and maximum as 8.30 during January-May.

Dissolved oxygen content. It ranges from 5.0 to 10.5 ppm, with maximum values during January-February, and minimum values during monsoon months, *i.e.*, during July-September.

Free carbon dioxide. There is a significant fluctuation on the free carbon dioxide level, being 0.6 ppm at Kanpur and as much as 10.0 ppm at Patna and Bhagalpur. It is invariably present in all places during July-November, being minimum during November-December and maximum during June-September.

Carbonate and bicarbonate contents. They range from 1.0 to 12.0 ppm. They are significantly present only during December-June. However, bicarbonates are present throughout the year, being in ranges from 72.0 to 212.0 ppm. The minimum values occur during July-September and the maximum during December-April.

Chloride contents. They show fluctuation from 4.0 to 35.4 ppm, being minimum during July to August. There is a steady rise during May-June.

Phosphates. They range from 0.05 to 0.21 ppm, the maximum values being during early monsoon, *i.e.*, in the month of July.

Nitrates. They show fluctuation from 0.08 to 0.22 ppm during different parts of the year.

Silicates. They fluctuate from 4.0 to 20.3 ppm, with maximum values during monsoon, or early winter months.

Plankton Production

Plankton production shows a bimodal cycle, having one peak in June and the other peak sometime during December-March. However, July-October months show poor production of planktons. The peaks of phyto- and zooplanktons show coincidence in their appearance. The population of phytoplankton is poor during monsoon and autumn months, but in the months of May-June and December-March, the latter represented the period of secondary abundance. The dominant groups of phytoplanktons are as follows :

Bacillariophyceae. Asterionella, Amphora, Cyclotella, Cocconeis, Cymbella diatoma, Fragilaria, Gyrosigma, Gomphonema, Navicula, Nitzschia, Pleurosigma, Pinnularia, Synedra.

Chlorophyceae. Ankinastrum, Ankistrodesmus, Chorella, Closterium, Denticula, Desmidium, Eudorina, Hydrodictyon, Mongeotia, Pandorina, Pediastrum, Scenedemus, Spirogyra.

Myxophyceae. Arbaena, Cylindrospermum, Lyngbya, Microcystis, Nostoc, Oscillatoria, Phormidium, Rivularia.

Rotifers. The dominant groups of zooplankton are the rotifers, which are represented by Asplanchna, Brachionus, Conochilus, Filinia, Keratella, Monostyla, Notops, Notholca, Pedalion, Polyarthra, Rattulus, Rotaria, Synchaeta.

Bottom Fauna

The bottom fauna from river Ganges was studied by the research scientists of CIFRI (Allahabad Substation) during 1958-66 at seven centres, namely Kanpur, Allahabad, Varanasi, Ballia, Patna, Bhagalpur and Rajmahal. It was found that the benthic organisms were maximum at Kanpur and minimum at Ballia, being 3,476/m^2 and 71/m^2 in these places. The remaining 5 centres at Allahabad, Varanasi, Patna, Bhagalpur and Rajmahal showed average annual production (quantity of benthic organisms) as 1,121/m^2, 880/m^2, 1,593/m^2, 218/m^2 and 502/m^2 respectively. In these centres, the peak production was recorded in June, being 21,143/m^2 at Kanpur, 3,436/m^2 at Allahabad, 2,214/m^2 at Varanasi, 264/m^2 at Ballia, 473/m^2 at Bhagalpur and 1,859/m^2 at Rajmahal. Minimum concentration of benthic fauna was recorded during August-September in the upper stretches and during January-February in the lower stretches. Interesting enough, it was marked that insects were maximum at Kanpur (98.0%) and minimum at Patna (0.6%). Reverse to this, molluscs were maximum at Patna (99.4%) and minimum at Kanpur (1.9%). Annelids were maximum at Rajmahal (29.3%), but completely absent at Ballia and Patna.

Insects. Several kinds of insects are found in various centres of Uttar Pradesh and Bihar. Diptera larvae (Chironomids), caddiflies (Trichoptera), hemipterans and mayflies dominated in various centres. The order of dominance in various centres, as per reportings by several scientists are as follows :

Centre	Kinds of Insects
Kanpur-Allahabad	Diptera larvae (Chironomid)
Varanasi	Caddiflies (Trichoptera)
Bhagalpur	Hemipterans
Rajmahal	Mayflies

Insects occur in much smaller numbers at Ballia, Patna and Bhagalpur. The entire insect population consisted of hemipterans of various kinds.

Molluscs. Among molluscs, bivalves generally dominated in the upper stretches with gastrapod population in scanty numbers, or were absent. On the other hand, gastropods dominated the lower stretches with maximum concentration at Patna. Among gastropods, *Melanoides lineatus* (7,638/m^2 in September) and *Melania tuberculata* (1,179/m^2 in August) were most common. *Viviparous bengallensis*, *V. variatus, Indoplanorbis exustus* and *Digonistoma pulchella* were also in large numbers. At Bhagalpur, *Pila globosa* were in large numbers but *V. benglaensis* and *I. exustus* were also commonly found. The last two usually existed in Rajmahal collections also. At Varanasi, bivalves (most common species being *Corbicula striatella*) dominated the molluscan population, but Parrysia sp. were usually found in dominating numbers at Ballia.

Annelids. They were found in large numbers at Varanasi, Allahabad, Bhagalpur and Rajmahal, with maximum population at Varanasi. The Annelids were rarely found at Kanpur, Ballia, and Patna. However, oligochaetes were available at Bhagalpur during monsoon months with a concentration of 57/m^2 as maximum (Ray *et al.*, 1966).

River Yamuna

As river Yamuna is deeper than Ganga, retaining several times more water, there is conspicuous difference in the physico-chemical features of the two. The Ganga water spreads about

3.2-4.8 km during monsoon, but Yamuna river spreads only 0.8 to 1.6 km laterally during monsoon. The latter (Yamuna river) has feeble current during summer but a fast one during monsoon. Ganga river, on the other hand, has a reverse order of current. As river Yamuna has a clayey bed, it is rich in organic matters and aquatic vegetation (Chakrabarty *et al.,* 1959). Reverse to this, Ganga has a sandy bed and the population of aquatic vegetation is poor.

The physico-chemical features have been found to be as follows :

Factors	Quantity
Water temperature	19.2-32°C
Turbidity	2400 ppm in September, but clear from November onward.
Dissolved oxygen	3.1-3.5 ppm
Total alkalinity	85-270 ppm
Hardness	35-185
Chlorides	6-56
Nitrates	Traces to 0.50 ppm
Phosphates	0.03-0.09 ppm
Silica	22 ppm (on average)

Phytoplanktons were studied on the basis of quantitative tests of the samples of collected water. The findings have been as follows :

Phytoplankton : Its maximum density was 619 units in June but lowest (19 units) in October.

Zooplankton : Its maximum density was 189 units per litre in June, but lowest (9 units) in October.

The ratio of zooplankton to Phytoplankton was 1:3. Diatoms represented the dominant group (49.7%) among the phytoplankton and rotifers (53.6%) among the zooplanktons.

Bottom fauna. The bottom fauna of river Yamuna was studied by Ray *et al.* (1966) at two centres : Sujawan and Minto

Park Ghat at 19.2 km and 3.2 km upstream of the confluence with Ganga at Allahabad. It was found that chironomid larvae 40.27%, mollusca (20.30%) annelids (20.30%), amphipods (2.30%), other organisms (8.83%) were in noticeable numbers in Sujawan site, but Minto Park Ghat area had representation of these items in 79.20%, 1.88%, 3.53%, 19.59% and 0.80% levels. Chironomid larvae and annelid worms formed the dominant groups during the major parts of the year.

River Gandak

Examination of water of river Gandak at 6 different centres in a 184 kms stretch by David (1963) revealed the following features :

Parameters	Ranges
Temperature	19.5-22.9°C
Dissolved Oxygen	6.2-7.4 ppm
Turbidity	Less than 100
Conductivity	215-437 mhos
Alkalinity	84-220 ppm
Chloride	40-8.4 ppm
Nitrate	0.04 to 0.16 ppm
Phosphates	Traces to 0.18 ppm
Sio_2	0.44-2.00 ppm
Calcium	29.5-78.7 ppm
Iron	Traces to 0.32
Hardness	92-136 ppm

Brahmaputra River

The water of this river was examined at 16 centres in a stretch of about 640 km from Dibrugarh to Dhubri during February-March in 1959. The water was found alkaline at all centres. Other features were, as per findings by Motwani (1959) and Motwani *et al.* (1962), the following :

Parameter	Names of Centres : Average data Tezpur	Guwahati	Dhubri
pH	7.5	7.6	6.9
Specific conductivity (u mhos/cm at 25°C)	0.15	0.17	0.17
Organic carbon	0.26	0.26	nil
Available P_2O_5 (kg/ha)	42	38	35
Available K_2O (kg/ha)	115	195	60
Texture (field tests)	Sandy loam	Sandy loam	Sandy

Thus it was found that the river Brahmaputra was rich in inorganic contents like nitrate, silicate, phosphates and iron. The average monthly plankton directly varied from 14 to 18 units/litre at different centres. Among phytoplankton *Chlorophyceae*, *Bacillariophyceae* were found in abundance. Among these, spirogyra, ulothrix, Tabellaria, Zygnema, Gomphonema, Navicula, Oscillatoria were most commonly found. However, among zooplankton, cyclops, Bosmina, Daphnia, Filinia, Lecane, Brachionus and Copepod nauplii were in dominating numbers.

East Coast River System

River Godavari. The physico-chemical features of river Godavari were recorded by Chacko and Srinivasan (1955) at Dawlaiswaram Anicut in the year 1952. The range of changes in various parameters have been given in tabulated form below (for 9 months only) to indicate trends for the whole year.

Factors	Monthwise quantum Feb.	Mar.	May	June	Jul.	Aug.	Sept.	Nov.	Dec.
Temperature	27.5	31.9	36.4	29.3	30.4	31.2	—	29.0	27.9
pH	7.5	8.2	8.3	7.3	7.3	8.1	8.2	8.2	7.2
Dissolved Oxygen (ppm)	3.63	1.47	1.4	4.45	1.26	1.82	3.63	3.4	18.2
Free CO_2 (ppm)	6.6	Nil	Nil	24.0	6.6	Nil	Nil	Nil	3.4
Carbonates (ppm)	Nil	6.1	5.5	Nil	Nil	12.0	12.0	12.0	Nil
Bicarbonates	186.1	192.2	158.6	152.5	131.2	45.8	137.3	152.5	131.2

It was found that the oxygen content of the river was generally low in March-April and May (1.40 to 2.16), when the flow of water was very slow. The maximum value of 4.45 ppm was in the month of June, coinciding with the onset of flood. Free CO_2 was conspicuously present only during February, June, July and December.

Planktons. The volume of plankton was in highest level in July, which coincided with the inflow of flood water. Phytoplanktonic fauna were represented mainly by various diatoms, green algae and blue-green algae. Specific planktons were as follows :

Diatoms. Navicula, Cymbella, Gyrosigms, Diatoma, Pinnularia, Amphora, Nitzschia, Fragilaria and Cyclotella.

Green algae. Spirogyra, Zygnema, ulothrix, microspora, volvox and pediastrum.

Blue green algae. Oscillatoria, Merismopedia and Nostoc.

Zooplanktons. They are represented by Diaptomus, Pseudodiaptomus, Cyclops, Diaphanosoma, Daphnia, Cariadaphnia, Moina, Keratella and Rotaria.

River Krishna

Chacko and Srinivasan (1955) made hydrobiological observations of this river and found the parameters in following ranges :

Factors	Range
Temperature	
Oxygen (mg/1)	3.07-5.46
Carbonate ppm	7.8-31.0
Bicarbonate ppm	81.9-105.4
Chloride content ppm	10-30
Phosphates ppm	0.24-0.51

Planktons. The planktons were represented by copepods (*e.g.*, Cyclops, Diaptomus), Cladocerans (*e.g.*, Daphnia, Diaphanosoma, Moina), and rotifers (*e.g.*, Brachionus, Salpina).

River Kavery

Chacko and Srinivasan (1955) made elaborate studies of the hydrobiological conditions of the river Kavery near Bhavani town. They marked that the river received floodwater between June and October, when the water turned brown in colour. The ranges of various parameters are as given below :

Factors	Ranges (with additional remarks)
Temperature	26.0°C in January 30.9 in May 26.2-26.9 during July-August
pH	8.3 in May to 8.5 in February, but it dropped to 7.9 in June, 7.6 in July after onset of flood.
Oxygen concentration	1.47 in May to 5.65 ppm in February 1.00 ppm.
Free CO_2	1.00 ppm in Dec., nil in Jan.-May to 5.4 ppm.
Chloride	12.4 ppm in Nov. 15.0 in Dec.-Jan. to 89.5 ppm in June.
Bicarbonate	366.0 ppm in May, 183.0 ppm in June to 112.9 ppm in February and Nil in October.

Planktons. They were represented by filamentous algae (*e.g.*, Mongeolia, Spirogyra, Oscillatoria and Aphanocapsa), diatoms (Cymbella, Cyclotella, Synedra, Nitzschia, Navicula), Copepoda (*e.g.*, Mesocyclops, Diaptomus), Cladocera (*e.g.*, Ceriodaphnia) and Rotifera (Brachionus).

West Coast River System

River Narmada. The record of investigation of river Narmada for hydrobiological features by the fishery experts of Allahabad substation of the CIFRI in 1964 at Sisodra showed the following ranges of four factors :

Factors	Ranges of quantum		
	July	August	September
Turbidity	440-5,000	307-1500	321-403
Water temperature	26.5-30.5	27.0-30.0	27.5-9.0
pH	7.5-8.5	8.5	8.5
Dissolved oxygen (mg/1)	5.72-7.48	4.84-7.43	6.38-7.92

Planktons. Average plankton concentrations per litre varied from 0.06 to 1.14 in July, 0.38 to 0.6 in August and 0.72 in September.

The informations on various parameters of different sources of water have undergone noticeable alterations due to impact of various kinds of pollutants produced by industries, detergents, biocides, etc. in recent decades. The impact of the changed hydrobiological features on prospects of fish yields has been considerable in affected areas of natural water sources, as elaborated in the chapter, 'Pollution Hazards'.

28

Riverine Fisheries

India has emerged as the second largest producer of inland fish in the world. With the annual inland production of 4.5 million tonnes and marine production of 3.9 million tonnes, position of India in the world level has improved to attain seventh rank, that is, it has become 7th largest producer of fish in the world.

One of the strategies adapted for the development of inland fisheries has been the better management of the inland water sources through Fish Farmers Development Agencies (FFDAs) and invention of new techniques for catching fishes. At present there are about 400 FFDA, which have exerted efforts to enhance productivity in inland water sources and bring more inland water areas under their control for gross development to yield more fishes. As per latest recorded information, the inland water sources in the form of rivers and canals are spread in 1,71,334 kilometre stretch in the various parts of India. The break-up figure of stretch of rivers in different states and union territories is shown in a tabulated form ahead.

India is full of rivers and rivulets. The main rivers of India are Ganga, Yamuna, Brahmaputra, Mahananda, Godavari, Krishna, Cauvery, Narmada, Tapti and their tributaries as shown in figure before (Fig. 27.1). These rivers and their tributaries have very rich fish fauna and, thus, are good sources of economic gain to fish farmers and rich source of proteinous food to general consumers. Outline figures of main fish types, besides those shown in other chapters including chapter 2, *i.e.*, "Culturable Fishes" have been given ahead (Fig. 28.1 to 28.9).

Statewise Figure of Length of Rivers and Canals*
(As in 1995)

S. No.	States/UTs	Kms	S. No.	States/UTs	Kms
1.	Andhra Pradesh	11514	18.	Rajasthan	NA
2.	Assam	4820	19.	Sikkim	900
3.	Bihar	3200	20.	Tamil Nadu	7420
4.	Goa	250	21.	Tripura	1200
5.	Gujarat	3865	22.	Uttar Pradesh	31200
6.	Haryana	5000	23.	West Bengal (P)	2526
7.	Himachal Pradesh	3000	24.	Arunachal Pradesh	2000
8.	Jammu & Kashmir	27781	25.	Mizoram	1395
9.	Karnataka	9000	26.	Andaman and Nicobar	115
10.	Kerala	3092	27.	Chandigarh	2
11.	Madhya Pradesh	20661	28.	Delhi	150
12.	Maharashtra	1600	29.	Lakshadweep	NA
13.	Manipur	3360	30.	Pondicherry	247
14.	Meghalaya	5600	31.	Dadar and Nagar Haveli	54
15.	Nagaland	1600	32.	Daman and Diu	12
16.	Orissa	4500			
17.	Punjab	15270			
	Total (1 to 32)				171334

NA: Not available; P: Provisional

**Source*: Handbook of Fisheries Statistics, 1996, Govt. of India.

Fisheries of Ganga River System

Reports on arrival of fish at Agra and Allahabad on the river Yamuna and at Kanpur, Varanasi, Buxar, Ballia, Patna and Bhagalpur on the river Ganga during the year 1958-1966, as given by the scientists of Allahabad substation of CIFRI (Barrackpore) are as follows :

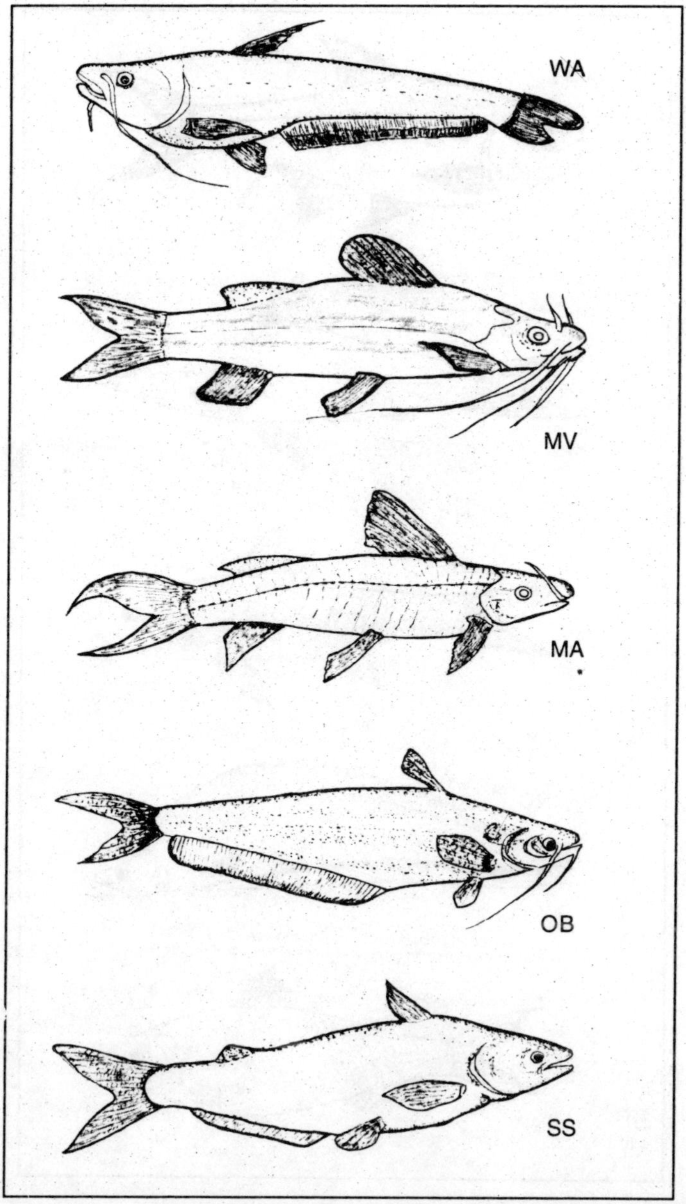

Fig. 28.1A, B. Outline figure of catfishes, *Wallago attu* (WA), *Mystus vittatus* (MV), M. aor (MA), *Ompok bimaculatus* (OB), *Silonia silondia* (SS).

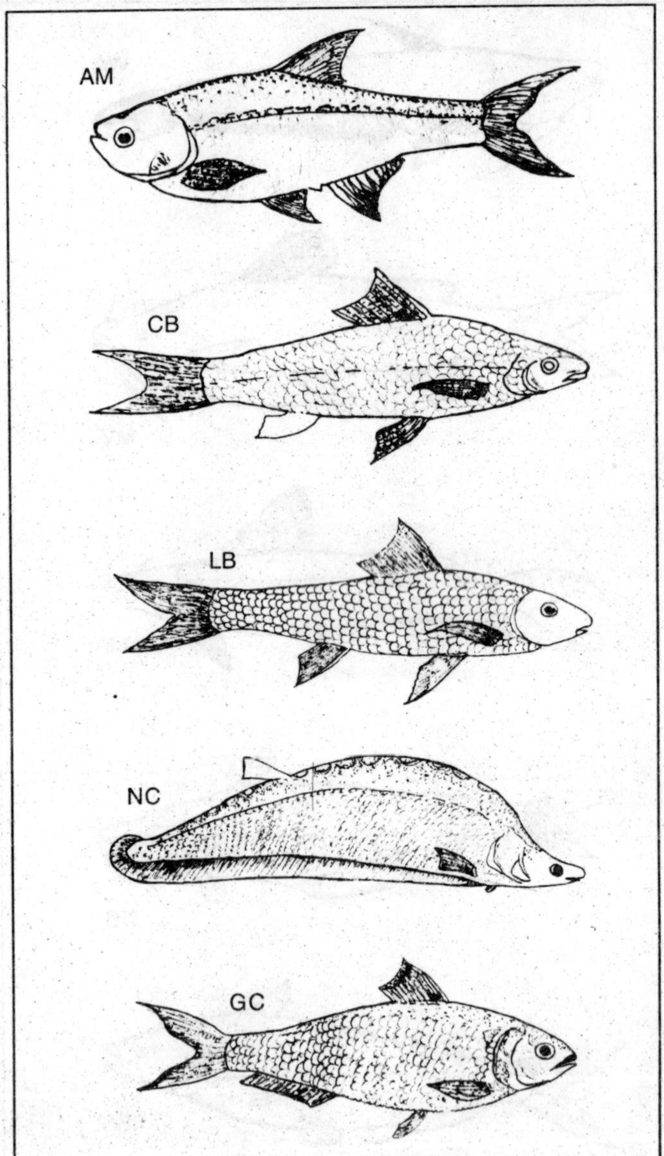

Fig. 28.2 to 28.5A. Outline figures of carps, *Amblypharyngodon mola* (AM), *Cirrhina reba* (CR), *Labeo bata* (LB); featherback, *Notopterus chitala* (NC); clupeid, *Gadusia chapra* (GC).

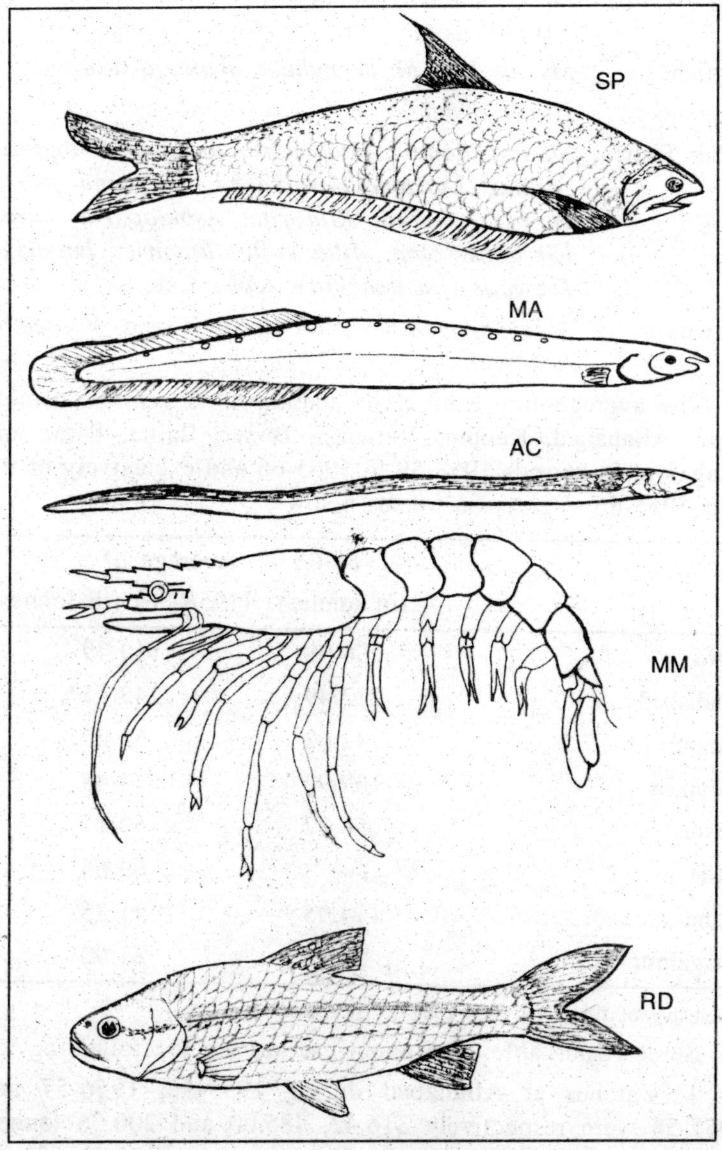

Fig. 28.5B to 28.9. Outline figures of clupeid *Setipinna phasa* (SP), spiny eel, *Mastacembalus armatus* (MA), eelfish, *Amphipnous cuchia* (AC), prawn, *Macrobranchium malcolmsonii* (MM) and hillstream fish, *Rasbora daniconius* (RD).

Major Carps : *Labeo rohita, L. calabasu, Catla catla and Cirrhinus mrigala.*

Catfishes : *Mystus aor, M. seenghala, Wallago attu.*

Clupeids : *Hilsa, ilisha.*

Other Fishes : *Silonia silondia, Pangasius pangasius, Notopterus chitala, Eutropichthys vacha, Setipinna phasa, Gadusis chapra, Rita rita, Aspidoparia morar, Cirrhinus reba, Ailia coila, Bagarius bagarius, Dasyatis (Pastinachus) stephen,* etc.

Prawns : *Macrobranchium malcolmsonii* and *Palaemon lamarrei.*

The average figure of total landings in 8 centres, namely Agra, Allahabad, Kanpur, Varanasi, Buxar, Ballia, Patna and Bhagalpur in periods 1958-59 to 1965-66 and exclusively of the year 1965-66 is given in tonnes below :

Centre	1965-66 (in tonnes)	Average of 5 Yrs.* (1958-66) (in tonnes)
Agra	60.08	119.49
Allahabad	212.49	243.25*
Kanpur	32.62	59.02
Varanasi	109.92	74.48
Buxar	56.12	59.17
Ballia	—	69.64
Patna	114.95	91.85
Bhagalpur	94.92	81.93

*Inclusive of figure of 1955-1958 (3 Years).

Some remarkable facts about catches are the following :

1. Catches at Allahabad during 1953-56, 1956-57 and 1957-58 were respectively 316.42, 485.00 and 200.06 tonnes. Among these, hilsa was contained to the extent of 35.7%, 62.6% and 29.7% respectively.

2. Major carps and some other selected catches (e.g., *Wallago attu*) formed the bulk of fish landings at various centres, *e.g.*, Allahabad, Varanasi, Buxar, Ballia, Patna and Bhagalpur.

3. *Cirrhinus mrigala* was most dominant species in upper stretches of Ganga from Kanpur to Allahabad and also in Yamuna. Average annual landings were found to be 36.01 tonnes each at Agra and Allahabad and 16.56 tonnes at Kanpur. At Varanasi, Buxar and Ballia, the annual landing was low, but a little better at Patna and Bhagalpur.

4. The representation of *Catla catla* was low throughout Ganga river system, being 16.86 tonne and 4.98 tonne at Allahabad and Kanpur, and in other centres its landing varied from 1.63 tonne to 4.99 tonne.

5. The landing of *Labeo rohita* was more in the upper and lower stretches of Ganga, being 14.82 tonne, 16.67 tonne and 16.15 tonne at Agra, Allahabad, but less in middle stretch, *i.e.*, at Varanasi, Buxar and Ballia. However, its abundance reappeared in Patna and Bhagalpur, being 7.59 tonne and 3.29 tonne as annual landing.

6. Annual landing figure of *Labeo calbasu* was sizable in the upper stretch of Ganga, *i.e.*, Agra and Allahabad centres, being 7.38 tonne and 9.57 tonne as annual landing in these two centres and 1.03 tonne at Kanpur, but in the middle and lower stretch was much less.

7. Annual figure of landing of *Mystus aor* and *Mystus seenghala* was maximum at Allahabad (34.28 tonne), followed by Agra (22.98 tonne) and Kanpur (9.98 tonne), but at other centres, there was less landing. However at Allahabad, this fish was next to mrigal in abundance (34.28 tonne).

8. *Wallago attu* had abundance in the upper stretch of the river Yamuna and Ganga. At Agra and Kanpur, the annual landing of this fish was 24.0 and 7.99 tonne respectively. The middle stretch of Ganga had level of landing, but at Bhagalpur, the average figure of annual landing reached 13.22 tonnes.

9. *Hilsa ilisha* had erratic annual figure of landing. The average annual landing was 20.16 tonne at Allahabad, but at Agra and Kanpur, the figure of landing was poor, but at Varanasi,

Buxar and Ballia, the annual landings were respectively 18.69, 22.56 and 29.26 tonnes respectively, forming 24.8%, 38.5% and 37.4% of the average annual landing of fishes.

The figure of annual landing of Hilsa at Allahabad reflected somewhat erratic trend, which may be related to the physical obstruction in the migration of hilsa in the upstream from Bay of Bengal. The stretch of Ganga river system has been shown in the figure before (Fig. 27.2).

The Brahmaputra River System

A survey work of fish fauna of this river system in Assam stretch by Motwani *et al.* (1962) revealed the presence of 126 species of fish belonging to 26 families. Out of these 41 represent fisheries of commercial importance. Out of 42 fish landing centres, Tezpur, Uzam Bazar and Fancy Bazar of Guwahati and Dhubri are four important main centres for recording statistics of fish landings. The average figure of fish landings in 1974-76 showed that catfishes dominated the figure of catch and the catch of major carps and prawns follow in descending order. Their respective percentage figures were 23.26% for catfishes, 17.85% for major carps, 14.26% for minor carps, 12.75% for hilsa, and 4.93% for prawns. Miscellaneous fishes represented 26.95%. Among different groups of fishcs, the following were main representatives:

Catfishes : *Wallago attu, Mystus montanus, Mystus vittatus, M. cavasius, M. menoda, M. aor, M. seenghala, M. bleekeri, Ompok bimaculatus, Rita rita, Silonia silondia, Heteropneustes fossils,Clarias batrachus, Eutropiichthys murius, Ailia ailia.*

Major Carps : Rohu (*Labeo rohita*), Dhawai (*Amblypharyngodon mola*), catla (*Catla catla*), mrigal (*Cirrhinus mrigala*), *Labeo gonius, L. calbasu* and *L. nandina.*

Minor Carps and other cyprinids : *Puntius conchonius, P. sophore, Ambassis ranga, A. nama, Labeo bata, L. mandiria, L. pangusia, Cirrhinus reba, Schizothorox* sp. *Ambassis ranga.*

Featherbacks : *Notopterus chitala; N. notopterus.*

Clupeids: Gadusia chapra, Hilsa ilisha, Setipinna phasa, G. variegata.

Murres : *Channa gachua, C. marulius, C. striatus, C. punctatus.*

Perches : *Anbas testudineus.*

Spiny eels : *Mastacembalus armatus, M. punctatus*

Eel fish : *Amphipnous cuchia*

Mahseer : *Tor tor, Tor putitora*

Hillstream fishes : *Noemacheilus botia, N. zonatus, N. scaturigina, N. savona, N. corica, N. beavani, N. sikmaeiensis, Rasbora diconius, Danio aequipinnatus, D. rerio, D. dovario, D. dangita, Glyptothorax telchitta, B. rebeiroi, B. striatus, Psilorhynchus baliotra, Balistra brucei, Semileptis gongota, Gangota nangra, Osteobrama cotio, Rasbora rasbora, Lepidocephalichtys berdmorei, L. annandalei, L. guntea.*

Miscellaneous : *Glossogobius giuris, Chela laubuca, Oxygaster bacaila, O. gara, Gagata cenia, G. viridescens.*

Estuarine fishes : *Setipinna phasa.*

Some remarkable features about fish landings in various centres of Brahmaputra river system are the following :

1. Among major carps, *Labeo rohita* and among catfishes, *Wallago attu* dominated the fish landings.

2. Hilsa was in abundant numbers in Fancy Bazar Centre (Guwahati) during all parts of the year, but in other centres the number was much less.

3. On examination of fishing gears, used for fish catches, the main gears or nets used were gill nets (with 12-250 mm mesh), drag nets (12-200 mm mesh), long lines, traps, and clap traps made of cotton/nylon/silk (Assam silk). The stretch of Brahmaputra river system has been shown in figure (Fig. 27.3).

The East Coast River Systems

The main rivers of east coast river systems are River Mahanadi, River Godavari, River Krishna, and River Kavery.

River Godavari : It has a length of 1,440 kms for its origin in the Deolali hills river near Nasik in the northern part of Western Ghat. It passes through regions of Maharashtra, Madhya

Pradesh, reaches Rajamundry in Andhra Pradesh and drains into the Bay of Bengal. It has a catchment area of about 3,51,980 sq. km. Its major tributaries are Manjira, Waingunga, Paingunga, Wardha, and Indiravati and the minor tributaries are the Purna, Maner and Sabari. Two large weirs have been built on river Godavari; one at Dowlaiswaram for irrigation and navigation, and the other at Dummagudem for navigation. No major Dam has been constructed on this river as yet. The stretch of the river system has been shown in Fig. 27.5.

Fisheries of river Godavari : For getting information on fish findings from river Godavari, the river can be divided into 3 zones. Zone I : It covered a 33.6 km long stretch between Dowlaiswaram and Patliseema, Zone II covered a 59.2 km long stretch between Polavanam and jidiguppa, and Zone III covered a 96.2 long stretch between Kunavaram and Dummagudem. 11 sampling centres consisting of 4 centres from zone I *i.e.*, Rajahmundry, Dowaiswaran, Bobbarlanka and Kovuur, 4 centres from zone II, *i.e.*, Polavaram, Divipatham Kondammodaler and Jidigappa, and 3 centres from zone III, *i.e.*, Kunavaram, Bhadrachalam and Dummagudem were recognised for getting information on fish landing. Figures of total catch in 1963, 1964, 1965, 1966, 1968, 1969 from these zones are given for 13 principal fishes and miscellaneous (mixed up) groups of fishes in Table 28.1.

The main fish group landed from river Godavari in the period 1963-1969, are the following :

Carps : *Labeo fimbriatus, Cirrhinus mrigala, Labeo calbasu, Catla catla.*

Catfishes : *Mystus seenghala, M. aor, Silonia childreni, Wallago attu, Pangasius pangasius, Bagarius bagarius.*

Hilsa : *Hilsa ilisha.*

Prawn : *Macrobranchium malcolmsonii.*

In respect to fisheries collected from river Godavari some remarkable observations are as follows :

1. It was found that *Labeo fimbriatus* was most commonly and abundantly available in all these three zones. Its annual landed figure was 35.1 tonne in 1963, but it came down to 5.9 tonnes

Table 28.1

Fishes	Yearwise figures of total catch						
	1963	1964	1965	1966	1967	1968	1969
Cirrhinus mrigala	19.3	18.6	18.5	18.5	12.6	16.4	11.8
Labeo fimbriatus	35.1	26.6	22.0	19.9	22.4	14.4	11.7
Catla catla	1.5	2.2	2.8	3.2	3.9	3.2	3.1
Labeo rohita	0.8	0.4	0.5	–	0.2	–	–
C. horai	4.8	6.9	1.4	–	1.1	–	–
Mystus seenghala	21.9	10.0	8.8	7.6	9.5	8.2	6.4
Wallago attu	6.1	3.0	2.0	2.1	2.8	2.5	4.2
Silonia childreni	4.1	4.4	3.0	2.9	2.9	1.4	1.6
Pangasius pangasius	5.2	3.2	1.6	–	1.1	–	0.5
Hilsa ilisha	15.5	46.4	6.3	17.9	14.9	21.7	21.2
Prawn	90.8	109.7	60.9	71.5	97.6	82.6	81.0
Labeo calbasu	7.8	9.7	5.7	4.8	7.6	7.5	5.5
Bagarius bagarius	7.8	4.9	2.9	–	2.1	→	–
Miscellaneous	109.3	75.8	69.2	82.7	83.1	75.6	71.0
Total	330.1	321.8	245.6	231.1	261.8	233.5	218.0

in 1969. This fall was nearly gradual (being in up position in 1968).

2. *Cirrhinus mrigala* was introduced in the Godavari river system from rivers of north India and it got established in course of time. Its total catch was 19.3 tonne in 1963, which came down to 16.4 tonne in 1968, and 11.8 tonne in 1969.

This fish represented carp fishery as next in importance to *L. fimbriatus*.

3. Among catfishes, *Mystus seenghala* was most important fish in quantity. Its figure of total landing varied from 21.9 tonne in 1963 to 6.4 tonne in 1969.

4. Prawn fishery was represented by one species, *Macrobranchium malcolmsonii*. Its annual figure of landing was 90.8 tonne in 1963, 109.7 in 1964, 97.6 tonne in 1967 and 81.0 in 1969. About 70% of commercial catch of prawns was in zone I at Rajahmundry.

5. Hilsa fishery was spectacular as its landing figure was in highest level as compared to figure of other fish kinds. Thus in 1963 it was 90.8 tonne and in 1964 it was 109.7 tonnes.

However, the trend declined irregularly, reaching 81.9 tonne in 1969.

The fishing gears used in river Godavari are generally of two kinds : (i) gill nets, which include set gill nets, drift gill nets, drag gill nets (bendu vala) and barrier gill net (katu vala), (ii) seines which include shore seine (Jaruga vala), large seine (alivi vala) and drag net (konte vala). In some places, cast nets are also used for fishing. Operation by gill net was most effective device as 33.9% of the total fishing gear effort was by this gear and 12.7% of the total catch was by this device. Next to the gill net operations was the drag gill net as effective device.

River Krishna

This river has its origin in the Western Ghat, and crosses a total length of 1,120 km. After crossing region south of Pune, passing through territories of Maharashtra, Karnataka and Andhra Pradesh, it reaches eastern coast and drains into Bay of Bengal. It has a catchment area of about 2,33,229 km. The main tributaries of Krishna river are the Bhima and the Tungabhadra rivers. The latter is formed by twin rivers, Tunga and Bhadra both of which originate in Mysore region. Some major reservoirs formed by high dams are the Lakkavalli, Tungabhadra, Koyna and Vanivilas sagar. Nevertheless, several reservoirs have been formed around Hyderabad city due to construction of dams access Musi rivers, Easi and some other rivulets. The main reservoirs thus formed are Himayat Sagar, Osman Sagar, Hussain Sagar, Meer-Alum tank and Ibrahim patar. A barrage has been constructed across the river Krishna at Vijayawada at a 80 kms distance from the mouth of this river.

Fisheries. David (1963) has reported that 91-136 kg of fish are captured daily from this river at Vijayawada. Details of fisheries from reservoir have been described for these reservoirs ahead in the chapter 'Reservoir Fisheries', and stretch of river Krishna has been shown in Fig. 27.5.

River Kavery

This river originates from the Brahmagiri Hills on the western Ghats at a height of 1.340 km, flows in southeast direction, drains into Bay of Bengal in Jhanjavur district of Tamil Nadu. It passes through rocky territory in Coorg, makes a fall at

Sivasamudram in the Karnataka state, and crosses Tamil Nadu state. The main tributaries of this river are Bhavni river, Noyit river and Amaravati river. A high dam has been constructed in the river Cauvery at Mettur and a big reservoir has been formed, commonly known as Mettur reservoir (also known as Stanley reservoir). It was completed in 1934.

Fisheries. Altogether 80 species of fish belonging to 23 families has been identified by Chacko *et al.* (1954) from river Cauvery (Kaveri). Among cyprinids the river representatives fish are *Aeronocheilus hexagonlepis, Tor patitora, Barbus carnaticus, Barbus dubins, Labeo kontius, L. ariza, Cirrhonus cirmoza, Osteochilus brevidorsalis* and *O. nashi. Among catfishes, the principal* representatives are *Glyptothorax madraspatanus, Mystus aor, M. seenghala, Pangasius pangasius, Wallago attu* and *Silonia silondia*, Mureels are represented by *Channa marulius* and feather-backs by *Notopterus notopterus*, Besides these some Gangetic carps such as *Catla catla, Labeo rohita, Cirrhinus mrigala* are also established here. Two exotic species, successfully transplanted from Cauvery river are *Cyprinus carpio* and *Osphronemus goramy*. For more informations on kinds of fisheries in this river system, the description given ahead for reservoir fisheries exclusively suffices the purpose.

The West Coast River System

This river system comprises two main rivers, Narmada and Tapti and their several tributaries. River Narmada originates in Bilaspur district of Madhya Pradesh at an elevation of 1,057 m in the Amarkantak hills. It has a total catchment area of 94,235 sq. km, which extends in the regions of Madhya Pradesh and Gujarat. It has 18 main tributaries of which 16 lie in Madhya Pradesh and remaining two in Gujarat. Reportings made by experts of CIFRI (Barrackpore, West Bengal) on the basis of landings in 1958-59 to 1965-66 at two representative centres, Hoshangabad and Shahganj are given here.

Fisheries. The annual landings ranged between 32.3 and 57.2 tonnes during 7 years, extending from 1958-59 to 1965-66, and the average yield was 41.5 tonne. The commercial catches of river Narmada comprised 23 species of fishes out of which 10 belonged to carp group, 8 to catfish group, 2 to murrel group (*Channa* sps.), 2 to Spiny eels (*Mastacembalus* spp.) and 1 to featherbacks (*Notopterus notopterus*). Carp fishes were represented by *Tor tor, Labeo fimbriatus, Labeo calbasu, Cirrhinus mrigala, Labeo bata, Labeo dyocheilus, Puntius sarana, Catla catla, Cirrhinus reba* and *Labeo gonius*. All these constituted 60.4% figure. Catfish group was represented by *Rita pavimentata, Mystus seenghala, Mystus aor, Wallago attu, Clupisoma garua, Ompok bimaculatus* and *Mystus cavasius*.

Among major carps, *Tor tor* (28.0%) and *Cirrhinus mrigala* (2.5%) had representation in the descending order as mentioned within brackets. Likewise, order of representation of minor carp with percentage figures (shown within brackets) was like this. *Labeo bata* (1.7%), *Puntius sarana* (1.4%) and *Cirrhinus reba* (0.5%). Catfish fishery was represented by *Rita pavimentata* (10.2%), *Mystus seenghala* (9.0%), *Wallago attu* (7.7%) and *Mystus aor* (4.7%), having order of representation shown within brackets. Other catfishes, *e.g.*, *Clupeisoma garua, Ompok bimaculatus* and *Mystus cavasius* had collective percentage figures of 2.5% among the total landings. Among miscellaneous groups, murrels (*Channa* spp.) had representation of 3.7%, *Mastacembatus* spp. (1.3%) and *Notopterus* spp. (0.6%).

River Tapti

Fishery season in river Tapati begins in September-October and it continues generally upto January-February. Report of fish landing during 1959-60 is available mostly for 728 km long stretch. The fish landings are represented mainly by *Tor tor, Labeo fimbriatus, Mystus seenghala, M. aor, Wallago attu, Labeo calbasu, L. bata* and *L. boggut*. Fingerlings of several commercially important fishes like *Tor tor, L. fimbriatus, Wallago attu, M. seenghala, M. aor, L. bata* and *L. boggut* are also available.

During winter and summer, the gears commonly used are cast nets, gill nets and long lines. Mahajal is also used in some areas like Khandwa, Jalgaon, Dhulia and Surat districts, but its overall popularity is less even now.